BAYES'S THEOREM

BAYES'S THEOREM

Edited by
RICHARD SWINBURNE

Published for THE BRITISH ACADEMY
by OXFORD UNIVERSITY PRESS

Oxford University Press, Great Clarendon Street, Oxford OX2 6DP

Oxford New York

Athens Auckland Bangkok Bogotá Buenos Aires Cape Town
Chennai Dar es Salaam Delhi Florence Hong Kong Istanbul Karachi
Kolkata Kuala Lumpur Madrid Melbourne Mexico City Mumbai Nairobi
Paris São Paulo Shanghai Singapore Taipei Tokyo Toronto Warsaw

with associated companies in Berlin Ibadan

Published in the United States
By Oxford University Press Inc., New York

British Library Cataloguing in Publication Data
Data available

ISBN 0–19–726267–8
ISSN 0068–1202

Typeset in Times
by Latimer Trend & Company Ltd., Plymouth
Printed in Great Britain
on acid-free paper by
Creative Print & Design, Ebbw Vale, Wales

Contents

Notes on Contributors   vii

Preface   ix

1. Introduction   1
   RICHARD SWINBURNE

2. Bayesianism — its Scope and Limits   21
   ELLIOTT SOBER

3. Bayesianism in Statistics   39
   COLIN HOWSON

4. Bayes's Theorem and Weighing Evidence by Juries   71
   A. P. DAWID

5. Bayes, Hume, Price, and Miracles   91
   JOHN EARMAN

6. Propensities May Satisfy Bayes's Theorem   111
   DAVID MILLER

Appendix   'An Essay Towards Solving a Problem in the   117
Doctrine of Chances' by Thomas Bayes, presented to the Royal
Society by Richard Price. Preceded by a historical introduction
by G. A. Barnard

Notes on Contributors

Philip Dawid is Pearson Professor of Statistics at University College London, and Past President of the International Society for Bayesian Analysis. He has an active interest, both theoretical and applied, in logical problems of legal reasoning.

John Earman is University Professor of History and Philosophy of Science at the University of Pittsburgh. He is the current President of the Philosophy of Science Association. His latest book is *Hume's Abject Failure: The Argument Against Miracles* (Oxford University Press, 2000).

Colin Howson is Professor of Philosophy at the London School of Economics and Political Science. His latest book is *Hume's Problem: Induction and the Justification of Belief* (Clarendon Press, 2000).

David Miller is Reader in Philosophy at the University of Warwick. In recent years he has held visiting appointments in Argentina, Brazil, and Mexico. From 1993 to 2001 he was Secretary of the British Logic Colloquium.

Elliott Sober is Hans Reichenbach Professor and Henry Vilas Research Professor of Philosophy at the University of Wisconsin; he is also a Centennial Professor at the London School of Economics and Political Science. His most recent book, co-authored with David Sloan Wilson, is *Unto Others: The Evolution and Psychology of Unselfish Behavior* (Harvard University Press, 1998).

Richard Swinburne is Nolloth Professor of the Philosophy of the Christian Religion in the University of Oxford and a Fellow of the British Academy. His latest book is *Epistemic Justification* (Clarendon Press, 2001).

Preface

THE INTRODUCTION TO THIS VOLUME and the four papers of Elliott Sober, Colin Howson, Philip Dawid, and John Earman are developed versions of papers given at a British Academy symposium on Bayes's Theorem on Saturday 10 March 2001. I have also added to the volume a short paper by David Miller dealing with the applicability of Bayes's Theorem to physical probability, an issue not explored in the symposium papers but important for a full treatment of the applicability of Bayes's Theorem. I have also added the original paper by Thomas Bayes which introduced his theorem to the world, including the introductory and concluding sections of that paper written by Richard Price when he presented the paper to the Royal Society on 23 December 1763. The volume is completed by a short historical introduction to the original paper, written by G. A. Barnard, originally published, together with the version of that paper printed here, in *Biometrika* **45** (1958): 293–315.

Richard Swinburne

1

Introduction*

RICHARD SWINBURNE

1. Kinds of probability

BAYES'S THEOREM is concerned with probability. When from the seventeenth century onward people began to talk about things being probable in somewhat like modern senses and reflected on what they meant, sometimes they supposed that there was only one kind of probability and sometimes they supposed that there were two kinds of probability — one a feature of the physical world, and the other the probability on evidence that something was the case in the physical world. Among modern philosophers the distinction was made most sharply and influentially by Rudolf Carnap, who called the former 'probability$_2$' and the latter 'probability$_1$'.[1] In my view Carnap seriously underestimated the number of kinds of probability, that is, the number of things which have usefully been called 'probability' in either ordinary language or technical literature, and which are susceptible to philosophical analysis and mathematical articulation.

Carnap's 'probability$_2$' is statistical probability. A statistical probability is simply a proportion in an actual class or in a hypothetical class, that is a class generatable by a repeatable process. Ordinary-language expressions in which the probability is said to concern 'a' member, that is, any member, of some class are naturally so interpreted. 'The probability of *an* inhabitant of New Hampshire in the year 2000 voting for the Republican presidential candidate', and 'the probability of *a* toss of this coin being heads' are naturally interpreted as assertions about actual proportions or proportions in a class generatable by some procedure, for example tossing the coin for some time. Where

* Some sections of this Introduction correspond closely to passages in my *Epistemic Justification*, Clarendon Press, 2001, where these issues are discussed at far greater length, and I put forward my own more definite views about them. I am grateful to Oxford University Press for permission to reuse this material.

[1] Rudolf Carnap, *Logical Foundations of Probability*, Chicago: Universty of Chicago Press, 1950, chapter 2.

Proceedings of the British Academy, **113**, 1–20. © The British Academy, 2002.

the proportion concerns an infinite class, there must normally (if the assertion is to have clear content) be an understanding of the order in which members of the class are taken; and the assertion is then to be read as claiming that when more and more members of the class are taken in that order, (eventually) the proportion diverges less and less from the stated value.[2] Where the proportion concerns a hypothetical class of concrete objects (whether finite or infinite), there must be an understanding of what must remain constant and what is allowed to vary (and how) when new members of the class are generated. What is the probability of a toss of an evenly balanced coin with a head and a tail landing heads? Suppose the world is deterministic; that is, which event occurs at a given time is fully determined by each previous state of the universe (within some interval of time). Then if we specify fully enough the conditions in which each toss is to be made — the exact angle of toss, distance above the table of the toss, momentum imparted to the coin, distribution of air molecules in the room etc., the probability will be 1 for many sets of conditions, and 0 for many other sets of conditions. Which of these sets of conditions are we talking about? Normally, in ordinary talk, neither. We suppose that these conditions may vary. But how? In the proportions in which, when people normally toss coins, they do actually vary. For example, if most tosses of recent years are made approximately equally often from within areas of equal angular width within a range of from 90° to the floor to 30°, we suppose a similar variation in tosses when talking of the proportion of heads resulting from an infinite series of tosses. And so on for all the other circumstances which affect the result. Given this ordinary context, then the probability of heads is 1/2. I shall represent the statistical probability of a B being A by $Pr(A|B)$. That there are statistical probabilities (at any rate for finite classes) is uncontentious. Detailed modern explications of statistical probability begin with John Venn's *Logic of Chance* (1866)[3] and continue through Richard von Mises's *Probability, Statistics and Truth* (1928)[4] and Hans Reichenbach's *The Theory of Probability* (1949).[5]

There is, however, another feature of the physical world which most of us believe to exist, and which is often called 'probability' but which Carnap did

---

[2] More precisely — to say that the proportion of As which are B in an infinite class of As (taken in a certain order) is p is to say that for every finite number $d > 0$, there is some number of As n_d such that for any $n > n_d$, where there are n As (that is, the first n As in the order), r of which are B, $p + d > r/n > p - d$. It follows from this definition that there will not be a probability of an A being B, for all infinite classes of As, but only for those in which there is such a limiting value p.

[3] London: John MacMillan and Co.

[4] Second revised English edition, London: George Allen and Unwin, 1957.

[5] Los Angeles: University of California Press, 1949.

not discuss in his main book on probability. Most of us think that the occurrence of events is predetermined by prior causes, either totally or to some limited extent; and talk of 'probability' is used to measure the extent to which
some outcome is predetermined to happen. I shall call this measure a measure of physical (or natural) probability. It is the kind of probability analysed
by a 'propensity' theory of probability, such as that of Karl Popper.[6] An event
having a probability of 1 means that it is predetermined to happen — that is,
physically necessary; an event having a probability of 0 means that it is predetermined not to happen — that is, physically impossible.[7] Intermediate values measure the extent of the bias in nature towards the event happening or
not happening. To say that the probability now of this atom of C_{14} decaying
within 5600 years is 1/2 is to say that given the whole state of the world now,
it is not predetermined whether or not the atom will decay within 5600 years,
but that nature has an equal propensity to cause decay and to cause no decay
within that time. Physical probability is relative to time — as the time at
which or by which the event is predicted to happen or not to happen, draws
near, so (if that probability is not 1 or 0) the probability of its occurrence may
change. Most of us think that the physical probability of almost all macroscopic events (at times very close to their occurrence or non-occurrence) is
very close to 1 or 0, but many of us think that because of the indeterminism
of quantum theory many microscopic events have (at such times) intermediate degrees of physical probability. (There are also, I must add, a few philosophers[8] who think, as did Hume, that there are statistical patterns in nature, but
not underlying propensities in individual causes which produce them. But

---

[6] See Karl Popper, 'The Propensity Interpretation of Probability', *British Journal for the Philosophy of Science* **10** (1960), 25–42. Propensity theory was also the subject of one of the two lectures, 'A World of Propensities: Two New Views of Causality' in Popper's *A World of Propensities*, Bristol: Thoemmes, 1990. Popper's discussion in this lecture gives the impression that he supposes that (even where no quantum effects are involved) the outcomes of most macroscopic events, such as the out-come of a certain toss of a coin, are produced by propensities of set-ups close in time to the outcome other than ones very close to 1 or 0. I see no good reason to suppose that. The outcome of each toss of the coin is normally virtually predetermined by the exact angle of toss, force imparted to the coin etc.; and so the propensity of the set-up to produce heads is either very close to 1 or very close to 0.

[7] See the Additional Note to this Introduction and especially its final paragraph for the difficulty in representing physical necessity by a physical probability of 1, and physical impossibility by a physi-cal probability of 0; and how it can be overcome.

[8] For example, David Lewis holds (on the whole, with occasional qualifications and doubts) that the 'chance' (his substitute for physical probability) of a particular event just is the statistical probability of an event of that kind over all time (the kind being picked out in terms of the categories used in the best scientific theory). See his 'A Subjectivist's Guide to Objective Chance' (in his *Philosophical Papers*, vol. II, Oxford: Oxford University Press, 1986) and the later 'Humean Supervenience Debugged' (*Mind*, **103** (1994), 473–90).

Humeanism is very much a minority view.) Given the existence of physical probability, there are a few logical entailments between it and statistical probability. For example, if the physical probability of every A being B at some time is 1, then the statistical probability of an A being B at that time will also be 1. There are only limited such entailments. For example, it does not follow deductively from the physical probability of each A being B having a certain particular value other than 1 or 0 that the statistical probability of an A being B will have any particular value at all. Even if the physical probability is 1/2, the very improbable may happen (even in the infinite sequence) and every A be B. However, plausibly, it will be very improbable (in the sense of logical probability to be delineated below) that this will happen, and we need an account of the criteria of logical probability which yields this result.

Carnap's 'probability$_1$' is a measure of the extent to which some proposition e (which may state some evidence) makes another proposition h (which may state some hypothesis) likely to be true. It has a value 1 when e makes h certain (that is, when given e, h is certainly true), 0 when e makes not-h (the negation of h) certain, and intermediate values as e gives intermediate degrees of support to h. The evidence may include evidence of statistical or physical probability; and the hypothesis may also concern statistical or physical probability. It is the probability of h on the total evidence available at some time, e, by which it is rational to be guided in our actions at that time. Most of us in our unphilosophical and unmathematical moments think that there are objective truths about whether such-and-such evidence makes such-and-such a hypothesis very probable, or only fairly probable or very improbable, though we doubt whether very precise values can often be given to this degree of probability. When this evidential (or epistemic) probability is understood as measuring the objectively correct degree of evidential support, I shall call it logical (or inductive) probability. Detailed modern explications of evidential probability began with J. M. Keynes's *A Treatise on Probability* (1921).[9] Keynes supposed that (at any rate often and approximately) there are true values for the extent to which one proposition makes another one probable, and so his account is an account of logical probability.[10] Keynes's work was developed by Rudolf Carnap, who sought in his *Logical Foundations of Probability* (1950) to give an explication of what was involved in what he regarded as an objective concept of 'evidential probability', his 'probability$_1$'.

---

[9] London: Macmillan.

[10] 'Logical probability' is not a fully satisfactory name for this kind of probability. I use it because it has been often used in the past to designate a priori objective evidential probability, but I do not assume that the value of every such probability (that is, of $P(h/e)$ for each h and e) is a 'truth of logic'. For there may be a priori truths which are not truths of logic.

Strangely (to my mind), many philosophers and statisticians in their philosophical and mathematical moments deny that there is such a thing as logical probability. Those of us who think otherwise allow that very precise values cannot often be given to the probabilities involved, that the probabilities take very rough values and that often all we can say truly is that this hypothesis is more probable than that one on such-and-such evidence, or that this hypothesis is more probable on this evidence than on that evidence. Those of us who believe that there is such a thing as logical probability usually think also that humans have roughly the same criteria as each other for assessing evidential support. We must, however, acknowledge that humans differ greatly in their ability to apply these criteria. Thus while the logical probability of h on e is 1 if e entails h, only someone who can recognize the entailment can see that the probability has that value. Many writers on probability, however, consider that there are no objectively correct standards for assessing the probability on evidence of particular hypotheses. Each person has their own standards; and so we cannot talk of logical probability but only of each person's subjective probability, which is a measure of the extent to which that person treats some evidence as supporting some hypothesis. Detailed explications of subjective probability, and the logical constraints to which rational measures of it are subject, began with the work of F. P. Ramsey (1926) and Bruno de Finetti (1930).[11]

2. Probability axioms

While there are these various kinds of probability, almost all writers consider that the same axioms (re-expressed as relating different kinds of entities — classes or propositions) govern (or in the case of subjective probability, ought to govern) both statistical and evidential probability (that is, both logical and subjective probability). These axioms were classically codified by Kolmogorov,[12] but were taken for granted or stated in more or less the same form for two or three centuries previously. As axioms of statistical probability, they can be stated as follows: for all classes A, B, and C:

1. $Pr(A|B) \geq 0$
2. If $B \subseteq A$, $Pr(A|B) = 1$

[11] For main papers by these two writers see H. E. Kyburg and H. E. Smokler (eds), *Studies in Subjective Probability*, New York: J. Wiley and Sons, 1964.

[12] A. N. Kolmogorov, *Foundations of Probability Theory* (first published in German, 1933), English edition, Chelsea Publishing Co., 1950.

3. If $A \cap B \cap C = 0$, $Pr(A \cup B/C) = Pr(A/C) + Pr(B/C)$
4. $Pr(A \cap B/C) = Pr(A/B \cap C) \times Pr(B/C)$
(3 and 4 do not apply if class C has no members.)

For finite classes, these are simple arithmetic truths; and they are naturally extended to the infinite domain.

Re-expressed as expressing relations between propositions, and $P(\ |\ )$ is an operator governing propositions, they become: for all propositions q, r, and s:

1. $P(q|r) \geq 0$
2. If $N(r \rightarrow q)$, $P(q|r) = 1$
3. If $N \sim (q \ \& \ r \ \& \ s)$ (that is, not all three can be true together),
   $P(q \ v \ r|s) = P(q|s) + P(r|s)$
4. $P(q \ \& \ r|s) = P(q|r \ \& \ s)P(r|s)$
(3 does not apply if $N \sim s$, that is if s is impossible.)[13]

Classes are individuated extensionally by their members, and so any two classes which have the same members are the same class. Propositions, however, are individuated by their intensions (what they mean) and not by the worlds in which they would be true. Thus 'the number of pebbles in my box is 3' and 'the number of pebbles in my box is $\sqrt{27}$' are different propositions, although they mutually entail each other and so are true in the same worlds. It follows from axioms 2 and 4 that if p and q are necessarily equivalent, for any r, $P(q|s) = P(r|s)$, but (in order to have a kind of probability which takes account of all necessary equivalences) we need to add

5. If $N(q \equiv r)$, $P(s|q) = P(s|r)$.

In order to interpret these axioms as axioms of physical probability, that is, propensity to cause, we need to confine the atomic propositions[14] to ones denoting time-indexed total world states (events), and read the 'N' (necessarily) as 'of physical necessity'. It is, however, disputed whether, even so, the resulting axiom system is satisfactory in view of the asymmetry of causation (that if Q is (part of) the cause of R, R is not (part of) the cause of Q), which I shall for simplicity's sake assume to hold in the form that effects must be

---

[13] See the Additional Note to this chapter.
[14] Atomic propositions are signified by individual lower-case letters, such as 'q' or 'r', denoting world states Q and R. Events which are not total world states can be denoted by disjunctions of atomic propositions.

later than their causes. For then if Q (denoted by 'q') is earlier than R (denoted by 'r'), $P(q|r)$ will always equal 0 (for R can have no propensity to cause Q) and so therefore — it follows from the axioms — will $P(r|q)$ (unless $P(q) = 0$, in which case $P(r|q)$ is undefined). So there cannot be any propensities at all! Hence, either we need a different axiom system for physical probability,[15] or we need a more carefully expressed semantics for physical probability, in which, for example, all probabilities are time-relative. On the latter interpretation $P(q|r)$ is the propensity at a particular time, say the present moment, for the world to develop in such a way that Q occurs, given that it has developed in such a way that R occurs. That propensity will exist whether Q is earlier or later than R. This way of interpreting the axioms is expounded by David Miller in Chapter 6 of this book.[16] There are different solutions of these kinds in the literature. There is, however, very widespread agreement[17] that the traditional axioms provide a satisfactory set for evidential probability (subjective or logical). For this purpose, the propositions may report anything whatsoever, and the 'N' is to be read as 'of logical necessity'. $P(h|e)$ is the probability of h given e. If we use the notation to designate subjective probability, we can, if required, add a subscript to indicate whose subjective probability is being considered. (Some of the other contributors to this volume have represented evidential probability as a relation between the states of affairs designated by propositions rather than as a relation between propositions, and so have expressed it in ways such as $P(Q|R)$ rather than $P(q|r)$.)

The normal technique for justifying the axioms, treated as axioms of subjective probability, is to show that in some way your desires are inevitably not going to be satisfied unless you assess the probabilities that actions will achieve your desired goals in ways that are consistent with the axioms of the calculus. The simplest case of this arises with betting. If you judge that the probability of an event E is p and are prepared to make any bet for or against p at the corresponding odds, for example so that if you bet £x that E will occur, you win £$[(1-p)/p]x$ if E occurs and lose £x if it fails to occur; and if you bet £x that E will not occur, you win £$[p/(1-p)]x$ if E does not occur but

---

[15] See the alternative calculus developed by Fetzer and Nute, presented in J. H. Fetzer, *Scientific Knowledge: Causation, Explanation and Corroboration*, Boston Studies in the Philosophy of Science, vol. 69, D. Reidel, 1981, pp. 283–6.

[16] For another statement of this way of interpreting the axioms in terms of physical probability, see C. S. I. McCurdy, 'Humphreys's Paradox and the Interpretation of Inverse Conditional Propensities', *Synthese* **108** (1996), 105–26.

[17] But not total agreement. There are writers who have proposed rival axiom systems for evidential support. L. J. Cohen, for example, has written three main books in defence of a rival axiom system. See, for example, his *The Probable and the Provable*, Oxford: Clarendon Press, 1977.

lose £x if E does occur, someone can always make a 'Dutch book' against you, that is, bet with you in such a way that you will inevitably lose money — unless your probability judgments are consistent with the axioms of the calculus. So, if contrary to the axioms of the calculus you judge that in a two-horse race (with no ties allowed) both horses have a probability of 2/3 of winning and so you bet £2 on each at 1–2, then you will inevitably lose £1 whatever the result of the race. This kind of argument has been extended so as to apply where people do not bet in a literal sense, but where the gains and losses resulting from their actions are measured in terms of their value to the agent rather than in monetary terms, so as to show that being guided by probabilities which do not conform to the constraints of the calculus will lead in various circumstances to inevitable loss. But, even if these extensions are successful, at most they seem to provide a constraint on the judgments of probability which a given agent should make at a given time. They seem to provide no reason why judgments of probability by a given agent at different times taken together should conform to the calculus. Subjective probability theorists have always had difficulty in justifying 'conditionalization' — moving from the conditional probability of h given e being p, when it is not known whether or not e, to the probability of h which ought to guide action, when e becomes known and forms the total available evidence, being p.

In any case, the only constraint to which subjective probability is subject is a coherency constraint of the kind described. Some of the craziest judgments of what makes what probable are consistent with the axioms of the calculus. It is perfectly compatible with the calculus, when having observed the sun rise innumerable times in the past (e) and given any other knowledge about the past (k), to judge that there is a 0.99 probability that it will never rise again (h) — to claim that $P(h|e \ \& \ k) = 0.99$. So long as you are careful not to ascribe a value other than 0.01 to the probability that the sun will rise again, and so values summing to no more than 0.01 to the probabilities that it will rise tomorrow, it will rise the next day, and so on, your attributions are fully coherent. That leads many of us to think that there are further constraints on attributions of value to probabilities other than conformity to the calculus, criteria for attributing correct values to some of the probabilities which will enable correct values to be attributed to other probabilities by means of the calculus. Such rules will yield the logical probabilities of propositions on various pieces of evidence as opposed to various people's differing subjective probabilities.

From where are we to get these criteria? From a study of many different examples of when we would regard a certain hypothesis as rendered very probable by certain evidence, less probable by other evidence, very improbable on yet other evidence, we can gradually distil criteria to which correct judgments should conform. We study the actual practice of scientists and historians and detectives, in those cases where we think that these experts made the right judgment. We then add innumerable further imaginary cases, and reflect on what would be right to say about which evidence renders which hypothesis how probable in each of them. Sometimes it will be clear that not merely certain comparative judgments but certain numerical judgments are correct. On the sole evidence that 500 out of 1000 tosses of this coin are heads, surely the probability that the next toss will be heads is 0.5. Not every investigator will agree in their judgments about all cases, but there is—I suggest—a very large measure of rough agreement about what makes what how probable. From all these examples we can extract general criteria for determining logical probability. These criteria will include conformity to the calculus—for we assess probability judgments which do not conform to the calculus as false, for reasons additional to those considered earlier.

3. Bayes's Theorem

In 1763 Richard Price presented to the Royal Society an edited version of the paper of Thomas Bayes, entitled 'An Essay Towards Solving a Problem in the Doctrine of Chances'.[18] Its principal result was interpreted by Price as providing 'a sure foundation for all our reasonings concerning past facts', that is, as a claim about evidential probability. It is on that form of it that subsequent theoretical interest has been focused and to that form that the contributors to this volume devote their attention. (However, as far as the text of Bayes's essay is concerned, he might as well or instead be thinking of his theorem as concerned with physical probability.) Taking for granted the axioms which I have called 1, 2, and 5, Bayes states explicitly the axioms which I have called 3 and 4, and thence derives his 'Proposition 5', which is the theorem in his text the closest (in my view)[19] to what has subsequently been called Bayes's Theorem. As stated by Bayes, it is concerned with the probability of a first event having happened given that a second event has happened. Denoting the first one by r and the second one by s, and our prior or background evidence by k, Bayes's Theorem then reads:

$$P(r|s \& k) = \frac{P(r \& s|k)}{P(s|k)}$$

[18] Reprinted in this volume as the Appendix.
[19] Though not in Colin Howson's view—see his paper in this volume, p. 39.

There is, however, given the axioms, no need for '*r*' to denote a first event and '*s*' a second one. Allowing them to be any propositions at all, and naming them '*h*' and '*e*' after their most usual application ('*h*' for hypothesis and '*e*' for new evidence) and using axiom 4 for the value of the probability of a conjunction, we get: (given that $P(e|k) \neq 0$)

$$P(h|e \ \& \ k) = \frac{P(e|h \ \& \ k) \, P(h|k)}{P(e|k)}$$

(where $P(e|k) = P(e|h \ \& \ k) \, P(h|k) + P(e|{\sim}h \ \& \ k) \, P({\sim}h|k)$).

It is this that I shall henceforward call Bayes's Theorem. If we interpret it as concerned with evidence (*e*) and hypothesis (*h*), it says that the posterior probability of *h* (its probability, given *e* and *k*) equals its predictive power ($P(e|h \ \& \ k)$, often — rather misleadingly — called the 'likelihood of *h*')[20] multiplied by its prior probability (its probability given *k* alone) divided by the prior probability of *e*. The formula in parentheses states that the prior probability of *e* is the sum of the probabilities of its occurrence in the different possible world states of *h* being true and ~*h* (not-*h*) being true. Most thinkers have no problem in regarding Bayes's Theorem as an acceptable theorem of subjective probability. The controversial issue is its range of application. Is there a 'logical probability' to which it applies; are there criteria for ascribing correct values to its terms? The view that there is such a logical probability, which governs the traditional axioms of the calculus and so Bayes's Theorem, is sometimes called objective Bayesianism. So are there such criteria? In all areas of inquiry, or only in some?

4. Bayesianism — subjective and objective

Any division of evidence between new evidence of observation (*e*) and background evidence (*k*) is arbitrary. But it is sometimes useful to make such a division in order to assess the probability of some hypothesis when we are taking certain aspects of the wider situation for granted. In considering a hypothesis (*h*) about how neon behaves at low temperatures, we may have both experimental evidence with respect to neon (*e*) and evidence about how other gases behave at low temperatures (*k*); and it may be helpful to call the latter 'background evidence' and see it as supporting the hypothesis about

---

[20] This terminology, deriving from R. A. Fisher, is widely current in statistical literature. I call it misleading because in ordinary language 'likelihood' means the same as 'probability'; and $P(e|h \ \& \ k)$ is the probability of *e*, not of *h*.

how neon behaves. But the crucial problems about logical probability are most evident if we suppose that we do not have any contingent evidence from a wider field about how things behave in some narrow field. In that case k is some mere tautology, and $P(h|k)$ and $P(e|k)$ may be called the intrinsic probabilities of h and e respectively.

Even in this situation there are some extremely obvious criteria for ascribing values which would be accepted by everyone. These typically concern the value of $P(e|h \ \& \ k)$, when h is some hypothesis purporting to explain e. For example, the logical probability of a particular event on the sole evidence of a statistical probability, $P(Bx|Ax \ \& \ Pr(B|A) = p)$, is p. (The logical probability that this coin will land heads, on the sole evidence that it is tossed and that 60 per cent of tosses of this coin so far have landed heads, is 0.6.) Also, on the sole evidence that there is a physical probability of p at t that x will be B at $(t+1)$, the logical probability on that evidence alone that x will be B at $(t+1)$, is also p.[21] But many doubt whether there are any objective rules for ascribing intrinsic probabilities. In the main text of Bayes's paper, when he discusses the example of a ball thrown on to a table or plane, he makes an explicit assumption of prior probability — that 'there shall be the same probability that [a ball] rests on any one equal part of the plane as another, and that it must necesarily rest somewhere upon it'. He provides at that place no criteria for reaching judgments of prior probability. But in the Scholium he tells us to calculate prior probability on the basis of a principle of indifference — which we may express, more precisely, as that we should assume (in the absence of experiment), for any variable about which we know nothing except that it lies between certain possible values, that it is equally probable that it will lie between any one interval of a given length as between any other interval of that length. But, as various paradoxes show, different ways of describing a set-up suggest different equiprobable ranges. Take the von Kries paradox, as stated by Keynes.[22] Suppose the specific volume of a liquid to lie between 1 and 3, and so its specific density to lie between 1 and 1/3. A principle of indifference applied to the former would suggest that the specific volume is equally likely to lie between 1 and 2 as between 2 and 3; and so — equivalently — that the specific density is equally likely to lie between 1 and 1/2 as between 1/2 and 1/3. Whereas, applied directly to the specific

---

[21] These two principles are two different ways of spelling out what David Lewis called 'The Principal Principle'. See his 'A Subjectivist's Guide to Objective Chance' in his *Philosophical Papers*, vol. II, Oxford: Oxford University Press, 1986.

[22] J. M. Keynes, *A Treatise on Probability*, p. 45. Sober gives a similar paradox in his paper in this volume, p. 21.

density, a principle of indifference would suggest that the specific density is equally likely to lie between 1 and 2/3, as between 2/3 and 1/3, which entails that it is not equally likely to lie between 1 and 1/2 as between 1/2 and 1/3. So which is the right way of allocating prior probabilities? And when the range of alternative hypotheses is not just hypotheses differing in respect of which intervals of the same variables they declare to be equiprobable (and so with a different distribution for the probability density), but much more sub-stantially — in terms of the kinds of entity and property they postulate and the mathematical relations the values of their properties have to each other — how on earth is any comparision of prior probability possible? Surely one can make no judgments of prior probability in advance of any evidence.

5. Prior probability and simplicity

The objectivist Bayesian answer is that if one could not make judgments of relative probability before evidence, one could not make them afterwards either. For compatible with any finite collection of data — both new observa-tional evidence and background evidence about what has happened in the past in a wider field — there is always an infinite number of incompatible hypothe-ses which yield those data with probability 1 (or any lesser probability you care to choose). (There is an infinite number of curves which pass through a finite number of points, and otherwise diverge wildly. And an infinite number of those curves will pass through any finite number of new points you choose.) So, if we want to say, as normally we do, that despite yielding the data with probability 1, some of these hypotheses are more probable than others, there must be a priori factors which are determining this. These are the factors that determine the prior probability of a hypothesis on no relevant evidence, which I call its intrinsic probability. In my view a full and careful analysis of the procedures of investigation, along the lines described earlier, will reveal that there are two kinds of factor — scope and simplicity. The greater scope a hypothesis has — the more entities about which a hypothesis makes claims and the more and more precise claims it makes about them — the lower (for a given degree of simplicity) is its prior probability. And the simpler a hypothesis is, the greater (for a given degree of scope) is its prior probability. In my view the simplicity of a hypothesis is a matter of it postu-lating few entities of few kinds, attributing to them few properties of few kinds, concerning properties more readily observable, fewness of laws, indi-vidual laws relating few variables, fewness of terms in equations stating a law and the mathematical primitiveness of the mathematical objects and relations utilized in laws.[23] To the extent to which a hypothesis is simpler on the balance of these various facets of simplicity it is, in my view, simpler overall and so (among hypotheses of equal scope) has greater intrinsic probability. By one property Q being more readily observable than another one R, I mean that the predicate 'R' designating R is introduced by a definition in terms (at least in part) of the predicate 'Q' designating Q but not vice versa; thus if 'grue' is introduced by a definition such as that an object is 'grue' at a time t iff it is green and t is before 2050, or blue and t is after 2050 (and 'green' is introduced by means of paradigm examples of objects which are green), then green is a property more readily observable than grue. If, on the other hand, 'grue' is introduced by paradigm examples of objects which are grue (and green is introduced in the same way), then since before 2050 'grue' and 'green' will have the same paradigm examples, these words will mean the same. The general force of this requirement is to lead us, other things being equal, to ascribe greater probability to hypotheses concerning properties such as 'green' and 'pointing to 7 on the dial' than to hypotheses concerning the physicist's properties of enthalpy, isospin or hypercharge. But other things are often not equal, and hypotheses concerning the latter properties become very probable because of their great predictive power in comparison with hypotheses concerning the former properties. When the simplest hypothesis fails to predict well, the next simple hypothesis often acquires the greater posterior probability.

There is an interesting and superficially much more precise and unified account of the simplicity of a hypothesis than the rough account which I have just given, and which has had some considerable influence among physicists, first put forward by R. J. Solomonoff.[24] He proposed giving hypotheses intrinsic Bayesian probabilities in terms of the reciprocal of the minimum number of computational symbols (in 'bits', 0s or 1s, that is) needed to express that hypothesis, called its 'string length'. To give an example of string length — a random string of 10^8 bits would have a string length of about 10^8 (because you cannot summarize the string by a short equation), whereas a string of 10^8 1s or a string of alternating 1s and 0s could be expressed by equations using far fewer bits. This account gives roughly the same results as does mine for which is the simpler of two hypotheses relating the same properties, where their relative simplicity turns on the simplicity of the mathematical equation relating these properties. But Solomonoff gives no rules for

---

[23] I give my own detailed account of simplicity in my *Epistemic Justification*, Oxford: Clarendon Press, 2001.

[24] R. J. Solomonoff, 'A Formal Theory of Inductive Inference', *Information and Control* **7** (1964), 1–22.

comparing hypotheses relating different properties, that is, hypotheses using different vocabularies. Thus the formulae $F = \text{Gmm}'/r^2$ for the gravitational force and $F = \alpha ee'/r^2$ for the electrostatic force have the same mathematical form but relate different physical variables. And so do 'all emerealds are green' and 'all emeralds are grue'. Solomonoff's account needs to be supplemented by some such criterion as my criterion of the greater simplicity of hypotheses relating more readily observable properties. A formula $x = y$ could hide a very complicated relationship if it needs a great number of observations and complicated extrapolation therefrom to detect the value of x.

I suggest that among hypotheses of equal scope equally able to yield the data, we judge those which better satisfy criteria such as mine or those of Solomonoff to have higher prior probability. We see that from the fact that we judge such hypotheses to be more probable than hypotheses which satisfy the criteria less well on evidence predicted by all of them. Such judgments concern comparative prior probability, but in some cases we can judge that some hypothesis has some exact probability on certain evidence, which in turn will enable us to make more precise the criteria of prior porbability (and—we may hope—help us to see what is the correct way to apply the principle of indifference in at least some of the paradoxical cases).

It is crucial to distinguish the sense of simplicity which I have been discussing, on which simplicity is a criterion for choice among theories of equal scope, from the sense of 'simplicity' in which a theory being simpler than another one just is it having greater scope than the other. It was Popper, more than anyone, who championed an understanding of this kind. He began by equating simplicity (of the kind that is relevant to epistemology) with degree of falsifiability. He wrote: 'The epistemological questions which arise in connection with the concept of simplicity can all be answered if we equate this concept with degree of falsifiability.'[25] He claimed that the 'empirical content'—in my terminology, the scope of a theory—increases with its degree of falsifiability'.[26] He compared[27] four theories of heavenly motion: 'all orbits of heavenly bodies are circles' (p), 'all orbits of planets are circles' (q), 'all orbits of heavenly bodies are ellipses' (r), and 'all orbits of planets are ellipses' (s). He claimed that since planets are only one kind of heavenly body, and circles are only one kind of ellipse, p ruled out more possible states of affairs than did the others, and q and r each ruled out more states of affairs than s. p was thus easier to falsify than, for example, q, because an observation

[25] K. R. Popper, *The Logic of Scientific Discovery*, London: Hutchinson, 1959, p. 140.
[26] Ibid., p. 113.
[27] Ibid., p. 122.

of any heavenly body, not just a planet, not moving in a circle would suffice to falsify it; and for that reason p told you more, had greater scope, than q. For similar reasons p had greater scope than r and s; q and r each had greater scope than s.

Now there may be a lot to be said for having theories simpler in this sense. Big claims are theoretically more important than small ones; and if they can be falsified easily, at any rate some progress can often be made. Theories with great scope are, however, as such, as I have already noted and as Popper boldly proclaimed, more likely to be false than theories with small scope. And there is no point in taking the trouble to falsify theories that are almost certainly false anyway. It is at this point that simplicity in a different sense comes in, as a criterion of probable truth. In my terminology a theory that a planet moves in a circle, for example, does not as such have greater simplicity than the theory that it moves in an ellipse; it just has fewer free parameters (a circle being an ellipse with a particular eccentricity, zero), and thus has greater scope. The theory that it moves in a circle, however, may well be simpler in my sense than the theory that it moves in an ellipse of a particular non-circular shape (where both have the same number of free parameters). The issue of how 'simplicity' should be understood arises in the papers in this volume of both Sober and Howson. Sober operates with an understanding of simplicity in terms of scope which has consequences for the relative simplicity of different theories somewhat similar to those that Popper reaches. Sober then points out (p. 37) that for 'nested models' where a proposition p entails a proposition q (because, for example, p gives constant values to free parameters of q) — it follows from the axioms of the calculus — p cannot have a greater probability than q on the same evidence. Howson points out (p. 57) that the interest in simplicity as a criterion of higher prior probability is an interest in it as a criterion for comparing incompatible (and so non-nested) hypotheses.

Some philosophers who are unwilling to allow that there are any general domain-indifferent criteria of prior probability hold (in effect) that within wide domains of which we have some experience, we know which hypotheses have some 'plausibility'. We know the kind of factors that might be at work, and so have a finite number of hypotheses with which we can operate. In such cases we may perhaps in effect assume all 'plausible' hypotheses to have equal prior probabilities, and then accumulate evidence which is more probable given some hypotheses than others and so, by Bayes's Theorem, will raise the posterior probabilities of the former over the latter. And if the number of hypotheses being considered is small, it might seem that it will not often greatly matter (within limits) how one allocates prior probabilities, since evidence in the form of a large collection of statistical data may quickly give a high posterior probability to a hypothesis whose prior probability is low. However, the nagging question remains as to the criteria by which our prior experience of the field leads us to select certain hypotheses and not others as having 'plausibility'.

It rather looks as if 'plausibility' is just another name for 'probability', and that—the claim is—contingent background evidence about a wide domain (*k*) gives different degrees of prior probability to different hypotheses about a narrow domain. But when we have used up all our contingent evidence (putting it into *k*), only a priori criteria can tell us which hypotheses (among the infinite number of logically possible ones compatible with and able to predict with equal probability what has happened in the wide domain) have which degrees of prior probability. Unless there are true intrinsic probabilities for the wide domain, there cannot be true contingent prior probabilities for the narrow domain, nor any true posterior probabilities at all. Either no scientific conclusion about which hypothesis or prediction is more probably true than any other, has any objective warrant; or there are correct a priori criteria of intrinsic probability.

This issue of how to determine prior probabilities affects the very simplest situation in statistics, where the experimenter is testing between two hypotheses: the hypothesis that some factor makes a difference to something and the hypothesis that it doesn't, e.g. that smoking increases the risk of cancer or that it doesn't. The investigator collects statistics of those who develop cancer among two large samples of the population—those who smoke and those who don't—and finds, say, that the proportion of those who develop cancer is much greater in the first sample. But maybe there is some other explanation of this than that smoking increases the risk of cancer. Maybe all in the first group are underfed, or have heart disease, factors known to be relevant to the occurrence of disease. But the two samples are chosen so that each sample has an equal proportion of those affected by these conditions. Yet there will always be an infinite number of properties other than being smokers, which all members of the first sample will have, and no members of the second sample will have—'living in houses numbered...' (followed by the numbers of the houses in which members of the first sample live), or 'being members of a sample chosen by such-and-such a process'. (Perhaps the process of creating the sample via random-number tables conduces to cancer.) And so on. Hypotheses that such factors are at work are thought implausible—and rightly so. But we have never tested hypotheses of innumerable such implausible kinds in this kind of domain. It looks as if our reasons for regarding them as implausible must come from domain-indifferent considerations, that is from a total world-view that includes the view that (in the absence of strong positive evidence to the contrary) it is very improbable that the kind of factor cited affects disease. And since a very complicated world-view which held that the factors at work in our particular domain are very different from those at work in other domains would be perfectly compatible with all our observations so far, my view is that it is only a priori criteria of prior probability including simplicity (in my sense) that can justify our preference for a world-view that certain kinds of factors are at work in all domains.

The issue of whether science needs or can have a priori criteria of prior probability is a central issue for the contributors to this volume, and in their different ways all the contributors to the original symposium — Sober, Howson, Dawid, and Earman — deny that science can have these criteria. They would, I think, all also hold that as a matter of fact most of us only take certain hypotheses seriously and then (either by giving them equal prior probabilities or by allowing empirical evidence to discriminate between them in some way not governed by the calculus) we can use the apparatus of the probability calculus to give them different degrees of posterior probability in the light of evidence (though Sober also claims — see below — that Akaike's Theorem provides an additional criterion for discrimination between hypotheses, though not by affecting prior probabilities). My own view expressed in this introduction, as well as the view of Solomonoff, is that there are a priori criteria of prior probability and these allow us to ascribe intrinsic probabilities to all hypotheses.

Elliott Sober claims (p. 24) that a proposition does not have a prior probability until we have 'empirical information' about 'the process at work' which will bring it about that the proposition is true or false; only in such circumstances can the Bayesian apparatus be applied. 'Empirical information' may lead us to hold that some hypotheses in the field are 'plausible' and others are not. But there are no rules for mapping this informal talk of plausibility onto formal talk of probability. In the absence of such empirical information, hypotheses can be compared only in respect of their 'likelihoods', that is, in respect of how probable it is that you would find the evidence you do given the different hypotheses; not 'all epistemological concepts that bear on empirical inquiry can be understood in terms of the probabilistic relationships described by Bayes's Theorem' (p. 21). Colin Howson thinks otherwise: he holds that the axioms of the calculus and so Bayes's Theorem govern all relations of epistemic support between propositions — but only in the form of a consistency constraint. They limit the values you can consistently give to one probability, given the values that you ascribe to other probabilities. With probability logic, as with deductive logic (p. 67), 'what you put in as a premise will be at least as fallible and conjectural as what you get out as a validly derived conclusion'. This is, of course, a form of subjective Bayesianism. He goes on to defend the Bayesian account of statistical inference against the non-Bayesian accounts of R. A. Fisher, and of J. Neyman and E. S. Pearson. (Colin Howson's detailed treatment of statistics inevitably makes his paper somewhat more technical than the other papers in this volume.) Sober claims that an important recent theorem of the probability calculus — Akaike's Theorem — does provide some rationale for preferring simpler theories (in his sense of theories with fewer adjustable parameters) to others; Howson denies the relevance of this theorem.

Philip Dawid and John Earman both take for granted Bayes's Theorem and assume that we can derive prior probabilities from empirical data, without discussing the extent to which a priori considerations enter into the derivation. They then show the consequences of applying Bayes's Theorem to two areas of inquiry. Philip Dawid considers how juries should use Bayes's Theorem to weigh evidence in criminal trials. John Earman considers the kind of evidence which would show that a miracle had probably occurred (in the sense of a 'violation' or 'non-repeatable exception' to a law of nature). He sets his answer in the context of the sophisticated eighteenth-century debate about weighing 'eyewitness testimony' to the occurrence of some event against 'uniform experience' that events of the kind in question do not happen — to which debate Bayes's eighteenth-century theorem provided a crucial contribution.

ADDITIONAL NOTE:
Countable Additivity and Probabilities of 1 and 0

Axiom 3 is known as the axiom of finite additivity. It may be generalized into what is known as the axiom of countable additivity, in order to allow for infinite disjunctions, to read:

$$P(\cup a_i | r) = \sum_i P(a_i | r),$$

where $\cup a_i$ is the proposition that a member of the set of propositions a_i is true, and $\sum_i P(a_i | r)$ is the sum of the probabilities of each of the propositions a_i on evidence r, no more than one of which can be true given r. This plausible axiom will, however, give rise to contradiction, unless we allow infinitesimal numbers. For consider an infinite set of propositions a_i each of which is equally probable and one of which must be true given r. Take r, for example, as the proposition that the length of my desk is some particular exact rational number of metres between 2 metres and 3 metres, e.g. exactly 2.00467 metres;

and let each a_i ascribe a different such value to the length. If we say that the prior probability of each such length has the same finite value greater than zero, however small, it will follow that the sum of an infinite number of such values will be infinite; and so it would follow from the Principle of Countable Additivity that the probability that the desk has a length between 1 and 2 metres is infinite. Yet there cannot be a probability greater than 1 (which by axiom 2 is the probability, on any evidence, of a tautology). But, if we attribute a value of 0 to the prior probability of each possible value of the length, the probability that the length will lie between 1 and 2 metres will (by the Principle of Countable Additivity) be 0 — contrary to what is stated by r.

No contradiction is generated, however, if we adopt a mathematics of infinitesimal numbers, in which there is an infinite number of such numbers greater than 0 but less than any finite number. Such a mathematics, called non-standard analysis, was developed by Abraham Robinson (see Abraham Robinson, *Non-Standard Analysis*, North-Holland Publishing Co., 1966). This allows us to attribute the same infinitesimal value to each of an infinite number of prior probabilities, which sum conjointly to 1. If we do not adopt non-standard analysis, not merely will we be unable to calculate the probability of an infinite disjunction from the probabilities of each of its disjuncts and conversely; but we shall still have a problem about what to say about the probability of each of an infinite number of equiprobable disjuncts, one only of which is true. If we attribute a finite value to it, then however small that value is, the Axiom of Finite Additivity will have the consequence that the probability of the disjunction of some very large finite number of disjuncts will again be greater than 1. So we would have to attribute to each disjunct the probability 0. But that would involve saying that such a disjunct was just as certainly false as a self-contradiction. That seems implausible. There is *some* chance of winning in a fair lottery with an infinite number of tickets! The use of infinitesimals allows us to distinguish between the probability of a proposition whose negation is entailed by its evidence, and one whose negation is not so entailed but which has a value less than any finite value greater than 0. The use of infinitesimals also allows us to distinguish between the probability of a proposition entailed by its evidence, and the probability of a proposition not entailed by its evidence, but which has a value greater than any finite number less than 1 (for example the proposition that my desk has a length between 2 and 3 metres other than 2.00467 metres). Without using infinitesimals, $P(h|e) = 1$ can only be read as 'e makes h certain' in a sense of 'certain' which fails to distinguish between the two cases.

The use of infinitesimals allows us to make an analogous distinction for physical probability. For example, consider an indeterministic world in which a point particle has an equal physical probability of landing at any of the infinite number of points on a screen. We need to distinguish the probability of its landing at a particular point from the physical impossibility of it landing at all. By arguments analogous to these for logical probability, the probability of the former will be less than any finite number. If we allow infinitesimals, we can say that it has an infinitesimal value, whereas what is physically impossible has a probability of 0.

2

Bayesianism — its Scope and Limits

ELLIOTT SOBER

1. The math and the philosophy (Bayes's Theorem ≠ Bayesianism)

BAYES'S THEOREM is a consequence of the definition of conditional probability. However, this way of putting things tends to conceal a proviso that needs to be recognized. What is true is that

$$P(H|O) = P(O|H)P(H)/P(O),$$

if each quantity mentioned in the theorem is well defined.

It is not inevitable that all propositions should have probabilities. That depends on what one means by probability, a point to which I'll return. The claim that all propositions have probabilities is a philosophical doctrine, not a theorem of mathematics. This is where Bayesianism begins and Bayes's Theorem leaves off. But there is more to Bayesianism that this. Bayesianism, in its strongest formulation, maintains not just that propositions have probabilities, but that all epistemological concepts that bear on empirical inquiry can be understood in terms of the probabilistic relationships described by Bayes's Theorem. Of course, more modest Bayesianisms also can be contemplated.

As an illustration of what Bayesianism amounts to, consider the continuing philosophical puzzlement over the epistemic significance of simplicity. Scientists and philosophers often maintain that simplicity or parsimony is relevant to evaluating the plausibility of hypotheses. The challenge to Bayesianism is to map this informal talk of plausibility onto formal talk of probability. More specifically, a double application of Bayes's Theorem yields the following comparative principle:

$$P(H_1|O) > P(H_2|O) \quad \text{if and only if} \quad P(O|H_1)P(H_1) > P(O|H_2)P(H_2).$$

Proceedings of the British Academy, **113**, 21–38. © The British Academy, 2002.

If 'more plausible' is interpreted to mean *higher posterior probability*, then
there are just two ingredients that Bayesianism gets to use in explaining what
makes one hypothesis more plausible than another. This means that if simplicity *does* influence plausibility, it must do so via the prior probabilities or
via the likelihoods.[1] If the relevance of simplicity cannot be accommodated
in one of these two ways, then either simplicity is epistemically irrelevant or
(strong) Bayesianism is mistaken.[2]

2. The usual objection — priors

The standard objection to Bayesianism is to my mind correct. It often does
not make sense to talk about propositions' having objective prior probabilities. This is especially clear in the case of hypotheses that attempt to specify
laws of nature. Newton's universal law of gravitation, when suitably supplemented with plausible background assumptions, can be said to confer probabilities on observations. But what does it mean to say that the law has a
probability in the light of those observations? More puzzling still is the idea
that it has a probability before any observations are taken into account. If God
chose the laws of nature by drawing slips of paper from an urn, it would make
sense to say that Newton's law has an objective prior. But no one believes this
process model, and nothing similar seems remotely plausible.

Bayesians used to reply to this challenge by trying to specify a sensible
version of the Principle of Indifference. This has turned out to be a dead end.
The problem is that there is no unique way to translate ignorance into an
assignment of priors. Consider my garden, which is a square plot of land that
is between 10 and 20 feet on each side. Based on this information, what is the
probability that the garden is between 10 and 15 feet on each side? It might
seem natural to say that every length between 10 and 20 feet has the same
probability (density), in which case the probability is 1/2 that each side is
between 10 and 15 feet. However, the information I gave you is equivalent to
saying that the garden has an area that is between 100 and 400 square feet.
This description makes it sound natural to say that every area between 100
and 400 square feet has the same probability, in which case the probability

---

[1] In this paper I use the terms 'likelihood' and 'likely' in the technical sense introduced by
R. A. Fisher — the likelihood of a hypothesis H in the light of observations O is the probability that
H confers on H, not the probability that O confers on H. H's likelihood is $P(O|H)$, while its probability
is $P(H|O)$.
[2] Bayesians need not argue that simplicity is always relevant just by way of influencing priors, or that
it is always relevant just by way of influencing likelihoods. See Sober (1990) for discussion.

is 1/2 that the area is between 100 and 250 square feet. However, this entails that the probability is 1/2 that the square is between 10 and $\sqrt{250} = 15.8$ feet on each side. Applying the principle simultaneously to length and to area leads to contradiction. If the principle is to apply to just one of these, which should it be? No satisfactory answer has ever been provided.

Most contemporary Bayesians have given up on objective Bayesianism, and have taken the subjective route. If Newton's law of gravitation does not have an objective prior probability, perhaps each agent has a subjective degree of belief in that proposition before the evidence starts to roll in. If point values cannot be assigned to these degrees of belief, perhaps they can be said to fall in reasonably well-defined intervals. There are interesting questions to be addressed here concerning this psychological hypothesis. But even supposing that agents have subjective degrees of belief, my problem with subjective Bayesianism is that subjective prior probabilities do not have probative force.

To explain what I mean by this, I want to examine the fairly standard evolutionary idea that the (near) universality of the genetic code is evidence that all organisms now alive trace back to a single common ancestor. Crick (1968) argued that the code now in use is a 'frozen accident'—which nucleotide triplet codes for which amino acid is functionally arbitrary. If this is right, it is clear why the universality of the code favours the hypothesis of one common ancestor over the hypothesis that current life traces back to twenty-seven separate start-ups. The evidence discriminates between the two hypotheses in this way because there is a likelihood inequality:

$$P(\text{the code is universal} \mid \text{one common ancestor})$$
$$> P(\text{the code is universal} \mid 27 \text{ mutually unrelated groups}).$$

This reasoning is grounded in objective (if not incontrovertible) considerations about the evolutionary process. What would be added to this if one specified one's subjective degrees of prior belief in the two hypotheses? People may have different feelings here. And even if people have the same feeling, I don't see why that common feeling is epistemologically relevant. If science is about the objective and public evaluation of hypotheses, these subjective feelings do not have scientific standing.[3] When scientists read research papers, they want information about the phenomena under study, not auto-biographical remarks about the authors of the study. A report of the author's

---

[3] It would be a different matter if one had an empirically well-confirmed theory that allowed one to say how often life can be expected to emerge from non-life in various environments, and how often whole phylogenetic trees can be expected to go extinct. A process theory of this kind would provide an objective basis for the prior probabilities. See Sober (1999) for discussion.

subjective posterior probabilities blends these two inputs together. This is why it would be better to expunge the subjective element and let the objective likelihoods speak for themselves.

I am not suggesting that we should avoid Bayesian thinking even in the privacy of our own homes. If you have a subjective degree of belief in a hypothesis, by all means use Bayes's Theorem to update that degree of belief as you obtain new evidence. For those of us who feel at a loss to say anything about the plausibility that many hypotheses have in the absence of evidence, this is an invitation we will want to decline. However, the most important point is that when opinions clash, the disagreement cannot be resolved by pointing to the fact that different agents have different subjective priors. If the disagreement boils down to this, the agents have simply agreed to disagree.

Consistent with this objection to (strong) Bayesianism, there remains an important domain of scientific problems in which Bayesianism is entirely legitimate. When the hypotheses under consideration describe possible outcomes of a chance process (Hacking, 1965; Edwards, 1972), it can make perfect sense to talk about objective prior and posterior probabilities. If you draw at random from a standard deck of cards, the probability that you'll draw the six of spades is 1/52. This is a 'valid prior', but not because it is obtained a priori from some version of the Principle of Indifference, and not because it reports your subjective degree of belief. The prior is legitimate because it is based on empirical information about the process at work.[4] There are many contexts in which Bayesianism has important applications — medical diagnosis and legal proceedings provide plenty of examples. My point is just that Bayesianism can't be the whole story about scientific inference.

3. The retreat to likelihoodism

As the example about the universality of the genetic code suggests, likelihoods are often more objective than prior probabilities. This makes it attractive to regard likelihood as an epistemology unto itself (Edwards, 1972; Royall, 1997). In doing so, one is changing the question one expects one's epistemology to answer. As Royall points out, likelihoods don't tell you what to believe or how to act or which hypotheses are probably true; they merely tell you how to compare the degree to which the evidence supports the various hypotheses you wish to consider.

---

[4] The prior probability is properly so called, not because it is a priori (it is not), but because it is in place prior to one's taking the new evidence into account.

Likelihoodism[5] is sometimes criticized for entailing that perfectly absurd hypotheses often have likelihoods that cannot be bettered. If you draw the six of spades from a deck of cards, the hypothesis that this was due to the intervention of an evil demon bent on having you draw that very card has a likelihood of unity, but few of us would regard this hypothesis as very plausible. Doesn't it sound strange to say that your drawing the six of spades supports the demon hypothesis more than it supports the hypothesis that the card was drawn at random from a normal deck? Yet this is precisely what likelihoodism asserts.

Whatever the merits of this objection, it is not something that a Bayesian should embrace. The reason is that Bayes's Theorem tells us that the observation of the six of spades *confirms* the demon hypothesis, in the sense that it raises its probability. This is the familiar point that when a hypothesis entails an observation, and the observational outcome was not certain to occur, and the hypothesis's prior probability is neither zero nor one,[6] the observation confirms. It is entirely consistent with this point that the probability of the demon hypothesis remains very low and the normal hypothesis's probability remains very high. But if confirmation concerns the diachronic question of how probabilities *change* rather than the synchronic question of what a probability's *absolute value* is, then Bayesians have to concede that the observation of the six of spades confirms the demon hypothesis. If so, they should not cast a jaundiced eye on the likelihoodist's claim about differential support.

Likelihoodists can and should admit that the demon hypothesis is implausible or absurd, notwithstanding the fact that it has a likelihood of unity relative to the single observation under consideration. It's just that likelihoodists decline to represent this epistemic judgement by assigning the hypothesis a probability. Likelihoodist epistemology is modest in its ambitions: support gets represented formally, but plausibility does not. It thereby contrasts with (strong) Bayesianism, which, as I've explained, aims to characterize *all* genuine epistemic concepts.

There's another Bayesian criticism of likelihoodism that I want to consider. This is the idea that likelihoods have epistemic significance only when they are used in the context of Bayes's Theorem. Bayesians sometimes present this claim as if it were obvious, but it isn't. Why is Edwards (1972, p. 100) wrong when he endorses Fisher's (1938) view that likelihood is a 'primitive

---

[5] By likelihoodism, I mean the comparative principle that O supports H_1 more than O supports H_2 if and only if $P(O|H_1) > P(O|H_2)$. It is a further claim that *degree* of differential support is measured by the likelihood *ratio*. Formulations of the Likelihood Principle often combine these two elements; see Forster and Sober (2002) for discussion.

[6] Since Bayesians usually reserve priors of 0 and 1 for tautologies and contradictions, I take it that they will want to assign the demon hypothesis an intermediate prior probability.

postulate' — it stands on its own and requires no more ultimate justification? Furthermore, it should be clear that the likelihood principle can be evaluated in the same way that any philosophical explication can be — by seeing if it accords with and systematizes intuitions about examples. Finally, I should add that there are non-Bayesian inferential frameworks in which likelihood plays a prominent role (Forster and Sober, 2002); I'll mention one of these at the end of this paper.

4. The trouble with likelihoods

Likelihoodism's objection to Bayesianism comes back to haunt it when one considers composite hypotheses. A simple statistical hypothesis confers a sharp probability on each possible observation. Composite hypotheses do not. When composite hypotheses have objective likelihoods, their values are often unknown; often they cannot be said to have objective likelihoods at all.

Composite hypotheses are frequently disjunctions — whether finite or infinite — of simple hypotheses. When this is so, their likelihoods are *averages*. Consider, for example, how the standard Mendelian model of reproduction may be used to compute the likelihoods of different hypotheses about the genotypes of an organism's parents, given an observation of the offspring's genotype:

$$P(\text{Offspring is } Aa \,|\, \text{Parents are } AA \text{ and } AA) = 0$$
$$P(\text{Offspring is } Aa \,|\, \text{Parents are } AA \text{ and } Aa) = 1/2$$
$$P(\text{Offspring is } Aa \,|\, \text{Parents are } AA \text{ and } aa) = 1.0$$
$$P(\text{Offspring is } Aa \,|\, \text{Parents are } Aa \text{ and } Aa) = 1/2$$
$$P(\text{Offspring is } Aa \,|\, \text{Parents are } Aa \text{ and } aa) = 1/2$$
$$P(\text{Offspring is } Aa \,|\, \text{Parents are } aa \text{ and } aa) = 0$$

Hypotheses about the genotype of the parental pair are simple. However, the hypothesis (H_1) that the offspring's mother is AA is composite; its likelihood is an average:

$$P(\text{Offspring is } Aa \,|\, \text{Mother is } AA)$$
$$= P(\text{Offspring is } Aa \,|\, \text{Mother is } AA \,\&\, \text{Father is } AA)$$
$$\times P(\text{Father is } AA \,|\, \text{Mother is } AA)$$
$$+ P(\text{Offspring is } Aa \,|\, \text{Mother is } AA \,\&\, \text{Father is } Aa)$$
$$\times P(\text{Father is } Aa \,|\, \text{Mother is } AA)$$
$$+ P(\text{Offspring is } Aa \,|\, \text{Mother is } AA \,\&\, \text{Father is } aa)$$
$$\times P(\text{Father is } aa \,|\, \text{Mother is } AA)$$
$$= (0)w_1 + (1/2)w_2 + (1)w_3 \qquad (\text{where } w_1 + w_2 + w_3 = 1.0).$$

The hypothesis H_1 that the mother was AA is a disjunction of simple hypotheses—either she was AA and the father was AA, *or* she was AA and the father was Aa, *or* she was AA and the father was aa. The same point holds with respect to the hypothesis (H_2) that the mother was a heterozygote. However, in this case the relevant simple hypotheses about the parental pair confer the same probability on the observation, namely 1/2. This means that H_2's likelihood is 1/2. The likelihoods of the two hypotheses, as a function of the father's genotype, are depicted in Figure 1.

Which hypothesis, H_1 or H_2, has the higher likelihood? To answer this question, we must evaluate the likelihood of H_1, but to do this we have to know the values of the weighting terms w_1, w_2, and w_3, which represent the properties of the mating scheme by which males and females in the parental generation came together to reproduce. If one had empirical information about whether mating was random or assortative (and to what degree), there would be no problem. But in the absence of such information, it is hard to see how values for these weighting terms can be specified unless one regards them as reflecting one's subjective degrees of belief.

In this example the weighting terms are 'nuisance parameters'. Our interest is in inferring the mother's genotype, but the father's genotype gets in the way. One solution that is sometimes employed in science is to estimate the values of nuisance parameters by finding the values for those parameters that maximize the likelihood of the composite hypothesis in question. In the case at hand, it is easy to see that $\hat{w}_1 = 0$, $\hat{w}_2 = 0$, and $\hat{w}_3 = 1.0$ are the values that maximize the

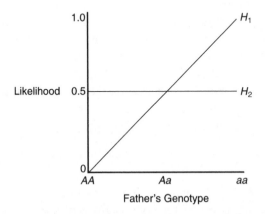

Figure 1. The likelihoods of two hypotheses about the mother's genotype, relative to the observation that the offspring's genotype is Aa. H_2 says that the mother was Aa; this hypothesis confers the same probability on the observation, regardless of what the father's genotype was. H_1 says that the mother was AA; what probability this hypothesis confers on the observation depends on the father's (unknown) genotype.

likelihood of H_1. If we assume that *AA* mothers always pair with *aa* fathers, we can conclude that H_1 has a likelihood of unity and so is more likely than H_2.

Although this procedure for handling nuisance parameters may seem to solve our problem, it does not. Rather, we have merely changed the subject. We have not compared H_1 and H_2; instead, we have compared $L(H_1)$ and H_2, where $L(H_1)$ specifies a specific set of values for the nuisance parameters in H_1. The likelihood of $L(H_1)$ is greater than the likelihood of H_1, not equal to it. H_1 is composite, but $L(H_1)$ is simple. Instead of assessing the *average* likelihood of the composite hypothesis with which we began, we have evaluated the likelihood of its likeliest special case.[7]

This example is artificial, but it illustrates a genuine issue that arises in scientific practice. As a more realistic example, consider the idea that the likelihood concept can be used to discriminate among competing phylogenetic hypotheses (first proposed by Edwards and Cavalli-Sforza, 1964, recently reviewed by Lewis, 1998). Suppose we are trying to ascertain the phylogenetic relationships that connect species *X*, *Y*, and *Z* and that our data consist of the characteristics these species are observed to exhibit. There are three possible phylogenetic trees — (*XY*)*Z*, *X*(*YZ*), and (*XZ*)*Y*. The first of these possibilities is depicted in Figure 2. The tips of the tree represent species that exist now; interior nodes represent common ancestors. The (*XY*)*Z* tree says that *X* and *Y* have a common ancestor that is not an ancestor of *Z*.

How can data on the similarities and differences these species exhibit be used to decide which genealogical hypothesis is best supported? The likelihood approach is to find the tree that confers the highest probability on the data. The problem, however, is that a phylogeny, by itself, does not tell us how probable it is that the three species should have the characteristics we observe. What is needed in addition is a quantitative model of how different traits evolve in different lineages. For example, the value of $P[\text{Data}|(XY)Z]$ depends on the rules of evolution that each trait obeys in the four branches depicted in Figure 2 and on the characteristics that the root species possesses. The likelihood of the tree topology is an average over the different possible values that these nuisance parameters can take:

$$P[\text{Data}|(XY)Z] = \sum_i P[\text{Data}|(XY)Z \,\&\, N_i]P[N_i|(XY)Z].$$

The hypothesis (*XY*)*Z* is composite — it is an infinite disjunction in which each disjunct consists of the topology (*XY*)*Z* with the nuisance parameters

---

[7] This 'best-case procedure' is discussed by Kalbfleisch and Sprott (1970), by Edwards (1972, pp. 109–19), and by Royall (1998, pp. 158–9).

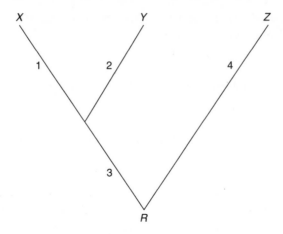

Figure 2. In this phylogenetic tree, there are nine nuisance parameters for each dichotomous character (whose two possible states are 0 and 1). For each lineage i ($i=1,2,3,4$), $e_i = P$(lineage i ends in state 1|lineage i begins in state 0) and $r_i = P$(lineage i ends in state 0|lineage i begins in state 1). In addition, there is one parameter that describes the state of the root — $P(R$ is in state 0).

fixed at a particular set of values. Biologists who use maximum likelihood typically estimate the values of nuisance parameters from the data. Thus, instead of comparing the likelihoods of $(XY)Z$, $X(YZ)$, and $(XZ)Y$, they compare the likelihoods of $L[(XY)Z]$, $L[X(YZ)]$, and $L[(XZ)Y]$. The problem has been changed from one that is intractable to one that can be solved.

A solution to this problem that is truer to the tenets of likelihoodism is to identify regions of parameter space where the likelihoods of the phylogenetic hypotheses have one ordering, and other regions where they have another. Perhaps when the nuisance parameters fall in one region, $(XY)Z$ is the hypothesis that makes the data most probable, whereas when the nuisance parameters fall in a different region, $X(YZ)$ is the hypothesis of maximum likelihood. This conditional assessment does not have the finality of the unqualified conclusion that $(XY)Z$ is the maximum likelihood hypothesis. Rather, it points to further biological questions that must be answered if we wish to say more (Sober, 1988).

The problem I have just surveyed — that composite hypotheses often do not have known objective likelihoods — is not a problem for likelihoodism, if that philosophy is sufficiently modest, but it *is* a problem for Bayesianism, if that philosophy is sufficiently *im*modest. Modest likelihoodists can and do admit that the likelihoods of composite hypotheses often cannot be evaluated. However, Bayesians who want their epistemology to provide a complete account of scientific inference here confront a second problem. The difficulty with priors also attaches to likelihoods.[8]

5. A further problem for the best-case strategy of dealing with nuisance parameters

Although Bayesians and likelihoodists should not confuse the likelihood of a composite hypothesis with the likelihood of its likeliest special case, it is worth exploring a further difficulty that arises if one changes the subject in this way. Let's return to the topology depicted in Figure 2 and consider what a fully realistic model of evolution in that tree will look like. We will want to have nine nuisance parameters for each dichotomous character. Eight of these are branch transition probabilities; the ninth assigns a probability to the root species' occupying character state 0. This model acknowledges that there can be *between-trait and within-trait heterogeneity*; different traits can evolve according to different rules, and the rules that a trait follows on one branch may differ from the rules that it follows on another. The model says that heterogeneities are possible, but it does not demand that they be actual. Two different parameters may have different values, but they also can have the same value. In addition to these nine parameters for each trait, a fully realistic model will also need parameters that represent the degree of independence with which each pair of traits evolves in each branch. The presence of these parameters in our model does not commit us to saying that traits are correlated in their evolution, but merely says that this is possible.

The complex model I have just described is realistic, but it has an embarrassing consequence — if we use this model and deploy the best-case solution to the problem of nuisance parameters, the result is that all phylogenies have a likelihood of unity. For example, consider a characteristic in which X is in state 1, Y is in state 1, and Z is in state 0; it is easy to find values for the nine nuisance parameters that pertain to this character's evolution that entail that this distribution of characters has a probability of unity. When we move to another character that has a different distribution, we can do the same thing. Other topologies are no different. What this means is that it is impossible to

---

[8] There is a Bayesian proposal for evaluating the average likelihoods of composite hypotheses. This is Schwarz's (1978) Bayesian information criterion (BIC). This approach imposes a flat distribution on parameter values that are near the data and a probability of zero on values that are far away; in addition, it renders commensurable the average likelihoods of composite hypotheses containing different adjustable parameters by introducing stipulations that fail to be invariant under reparameterization. See Forster and Sober (1994, pp. 23–4) for discussion.

discriminate among phylogenetic hypotheses if we use the best-case strategy in the context of a fully realistic model of the evolutionary process.

This has not stopped evolutionary biology in its tracks. Rather, biologists assign likelihoods to tree topologies by using constrained models. These constraints fall into two categories:

> *Within-trait constraints*: The strongest version of this idea is that a trait's probability of changing in a unit of time is the same everywhere in the tree. A weaker constraint is that a trait's probability of changing in a unit of time is the same at any two simultaneous temporal intervals. In terms of the topology depicted in Figure 2, the latter constraint entails that $e_1 = e_2$, but says nothing about the relationship of e_1 and e_3; the former says that $e_1 = e_2$ *and* that $e_1 = e_3$ if lineages 1 and 3 have the same durations.

> *Between-trait constraints*: Traits evolve independently of each other, and different traits in a lineage follow the same rules of evolution.

This second category of constraints has been applied both globally, to all traits in the data, and locally, within classes of traits.

Although these constraints save the problem of phylogenetic inference from collapsing under its own weight, they are manifestly unrealistic. Do we really believe that a trait's rules of evolution are exactly the same in different lineages? Do we really believe that different traits follow exactly the same rules of evolution? These are idealizations.[9] What is clearly true is the unconstrained model, which says that different traits may or may not evolve independently, that they may or may not follow the same rules of evolution, and that a given trait may or may not follow the same rules in different lineages. We thus have arrived at a dilemma: we can use a realistic model and give up on the idea of inferring phylogenies by best-case maximum likelihood, or we can use a constrained model to infer phylogenies, but leave our inference vulnerable to the charge that the model used is not realistic.

As noted earlier, the best-case strategy for dealing with nuisance parameters does not conform to the dictates of likelihoodism, which is perfectly clear on the difference between the average likelihood of a composite hypothesis H and the likelihood of the simple hypothesis $L(H)$. However, I doubt that this point will give pause to biologists who use this strategy to reconstruct phylogenies. This is because these biologists are frequentists, not likelihoodists; they use the frequentist technique known as the likelihood ratio test.[10] This test prevents the collapse I have just described. Instead of automatically opting for hypotheses of

---

[9] For discussion of the relationship between idealization and simplification, see Sober (1998, 2002) and Forster (2000, 2002).

[10] Don't be misled by the terminology — the likelihood ratio test is not consistent with likelihoodism.

Table 1

Process models	Tree topologies		
	(XY)Z	*X(YZ)*	*(XZ)Y*
All characters evolve independently (*N*90)	*L*[(*XY*)*Z* & *N*90]	*L*[*X*(*YZ*) & *N*90]	*L*[(*XZ*)*Y* & *N*90]
All characters evolve independently and follow the same rules (*N*9)	*L*[(*XY*)*Z* & *N*9]	*L*[*X*(*YZ*) & *N*9]	*L*[(*XZ*)*Y* & *N*9]
All characters evolve independently, follow the same rules, and evolve at a constant rate (*N*5)	*L*[(*XY*)*Z* & *N*5]	*L*[*X*(*YZ*) & *N*5]	*L*[(*XZ*)*Y* & *N*5]

high likelihood, they ask whether more complex hypotheses have likelihoods that are *significantly* greater than the likelihoods of simpler hypotheses.

Not that all is well if one embraces frequentism to solve this problem. In addition to the conceptual objections that Bayesians and likelihoodists have developed against frequentism, frequentism has a practical limitation in the problem at hand — the frequentist's likelihood ratio test applies only to nested hypotheses. This can be illustrated by considering our three competing tree topologies and some of the process models already described. Consider a data set that describes the character states of each species for ten dichotomous characters. Table 1 represents the different best-case hypotheses that are generated by bringing a tree topology together with a process model and then finding the maximum likelihood estimates of the nuisance parameters.[11] Models in the same column are nested, but entries in different columns are not.[12] Frequentism has no way to implement diagonal comparisons.

6. Simplicity — the Achilles Heel of (Strong) Bayesianism

In the list of process models displayed in Table 1, *N*5 is simpler than *N*9, and *N*9 is simpler than *N*90. Don't be misled by the lengths of the verbal descriptions; the relevant consideration is the number of adjustable parameters. Model *N*90 has 90 nuisance parameters — nine for each of the ten character distributions in

---

[11] In this table, I've written, for example, '*L*[(*XY*)*Z* & *N*9]' and not '(*XY*)*Z* & *L*[*N*(9)]'. The reason is that conjunctions in the same row often have their nuisance parameters set at different values, depending on the phylogeny to which they are attached.

[12] Although the conjunctions in this table that include the same phylogeny are nested, it is perfectly possible for two such conjunctions to be non-nested (e.g., let one assume that there is between-trait homogeneity and leave open whether there is within-trait homogeneity and let the other do the opposite).

the data set. Because it has so many parameters, this model is able to leave open whether different traits evolve according to the same or different rules, and also whether a given trait follows the same rules in different branches. In Table 1's list of models, simpler models entail models that are more complex. This is the kind of situation that Popper (1959) was thinking about when he equated simplicity with falsifiability.[13] As Popper observed, the epistemic relevance of simplicity in this instance cannot be captured by stipulating that simpler theories are more probable. If $N5$ entails $N90$, $N5$ cannot be more probable than $N90$.

Howson (1988) correctly notes that there is no logical prohibition against assigning simpler models higher priors when models are *not* nested. However, Popper's point remains true for nested models. What are we to conclude? It may seem that the question that needs to be addressed is whether scientists should compare nested models. As noted before, they in fact do so; scientists are often frequentists and the likelihood ratio test *requires* that models be nested. Bayesians may reply that this is a confusion from which scientists need to emancipate themselves. Since nested models are not incompatible, why should we regard them as competitors? I'll return to this question in a while; the point I want to emphasize here is that the insistence on non-nested models does not pluck the Bayesian fat from the fire. This is because it is obscure what justification the Bayesian can offer for assigning simpler models higher priors. For example, suppose we reformulate $N90$ so that $N9$ is no longer nested in it—let $N90^*$ use different parameters for different traits with the stipulation that different traits cannot have exactly the same branch transition probabilities. What justification could there be for assigning $N9$ a higher prior than $N90^*$? If a gun were put to my head, I'd allow dimensionality considerations to lead me to bet on just the opposite judgement— I'd say that it is more probable that two traits have different probabilities of evolving than that they have exactly the same probabilities of evolving.

If Bayesianism can't capture the epistemic relevance of simplicity by defending an objective ordering of prior probabilities, the other possibility is that it might be able to explain the relevance of simplicity via the vehicle of likelihoods. However, we have just seen that serious difficulties stand in the way of this undertaking. The likelihoods of composite hypotheses often cannot be evaluated objectively. And if we change the subject by using the best-case strategy, the problem is that simpler models inevitably come out with lower likelihoods, not higher ones.[14]

---

[13] Popper, of course, realized that simplicity is sometimes not epistemically relevant; he introduced his equation to explain what makes simplicity epistemically relevant when it is so in fact.

[14] For further discussion of Bayesianism's and likelihoodism's inadequate treatment of simplicity, see Forster and Sober (1994) and Forster (1995).

I mentioned at the outset that Bayesianism has just two resources for explaining the epistemic relevance of simplicity—priors and likelihoods. Neither of these appears to be at all promising. Does it follow that Bayesianism is mistaken? There is a way out to consider—perhaps one should deny that simplicity has any epistemic relevance at all. Perhaps simplicity is merely an aesthetic frill. Scientists *like* simpler theories for various reasons, but that does not mean that simplicity is epistemically significant.

I think this way out is blocked for two reasons. First, the practice of science makes it very hard to believe that simplicity always counts for nothing. It does no good for the Bayesian to point to examples of scientific inference in which simplicity plays no role. Granted, there are such cases. However, in biology and the social sciences, scientists frequently compare models that contain different numbers of adjustable parameters. Simplicity is central to science because model selection is a pervasive problem.

The second reason not to deny the epistemic import of simplicity is that there exists an inferential framework that is neither Bayesian, nor likelihoodist, nor frequentist, which entails that simplicity is epistemically relevant and explains why this is so. This is the model selection framework and criterion developed by Akaike (1973) and his school (see Sakamoto et al., 1986). It turns out that the simplicity of a model, when measured in terms of the number of adjustable parameters it contains, is relevant to estimating how predictively accurate the model will be. Akaike's framework and criterion are non-Bayesian, in that no prior probabilities are invoked. However, the criterion for model selection that Akaike derives does say that the likelihood of a model's likeliest special case is relevant to estimating the model's predictive accuracy. Akaike (1973) describes his proposal as an 'extension' of the method of maximum likelihood. Likelihood is relevant to estimating predictive accuracy, but it is not the only thing that is relevant; simplicity is relevant too.[15]

Since this paper is about Bayesianism, not the work of Akaike, I won't try to explain in any detail how these ideas work. However, I will make a few brief comments, which I hope will whet the reader's appetite. Akaike suggested that the problem of model selection be conceived in terms of a certain goal; the goal is not to find models that are true, but models that will be predictively accurate. This conception of the goal of model selection is what I mean by Akaike's 'framework'. Akaike also proposed a means for achieving that goal; he proved a theorem that describes how one can obtain an unbiased estimate of a model's predictive accuracy. This theorem is the basis for what

---

[15] For further discussion of Akaike's framework and theorem, see Burnham and Anderson (1998), McQuarrie and Tsai (1998), Forster and Sober (1994, 2002), Forster (2002), and Sober (2002a).

has come to be called the Akaike information criterion (AIC). This separation of Akaike's framework from his criterion is important; there may be circumstances in which AIC is *not* the best criterion to use in model selection, even granting the goal of maximizing predictive accuracy. The model selection literature contains a good deal of discussion of this point. My own view is not that AIC is the be-all and end-all; what I find philosophically interesting is the Akaike framework and a certain feature that many model selection criteria share—that simplicity is relevant because it helps one estimate predictive accuracy.

Akaike's idea of predictive accuracy has to be understood in terms of a two-step process. Models that contain adjustable parameters make predictions in the following sense: first one draws a set of data from the underlying distribution and uses those data to estimate the values of the model's parameters (by maximum likelihood estimation). One then uses that fitted model to predict a new data set drawn from the same distribution. In terms of our previous notation, we use a model M to make a prediction about new data by using the old data to find $L(M)$—it is $L(M)$ that makes a definite prediction. The predicted values may be close to the new data, or far away (as measured by the Kulback–Leibler distance measure). Imagine using the model repeatedly in this two-step process; there will doubtless be some variation among these repetitions in terms of how well the fitted model predicts new data. The *average* performance of the model is what defines its predictive accuracy. The predictive accuracy of M is the *expected* likelihood of $L(M)$.

Having models that are predictively accurate may be a desirable goal, but how can one tell how predictively accurate a model is apt to be? That is, is predictive accuracy epistemically accessible? Akaike's (1973) remarkable theorem provides an answer:

An unbiased estimate of the predictive accuracy of model $M \approx$ Log-likelihood$[L(M)] - k$.

One takes the logarithm of the likelihood of the fitted model and subtracts k, the number of adjustable parameters.[16] Complex models, when fitted to the data, tend to have higher likelihoods than simpler ones, but they also incur a larger penalty because of their complexity. For a complex model to have a higher AIC value than a simpler one, it isn't enough that the complex model fit the data better; it must fit the data better by a sufficient margin to overcome the fact that it is more complex.

---

[16] More exactly, the formulation of Akaike's result that Forster and Sober (1994) and Forster (2002) recommend is that an unbiased estimate of the model's predictive accuracy *per datum* is $(1/N)\{$Log-likelihood$[L(M)] - k\}$, where N is the number of data.

There is more to the Akaike framework than Akaike's Theorem. For example, even though AIC provides an unbiased estimate of a model's predictive accuracy, one may want to know how much error there is in this estimate. Sakamoto et al. (1986) describe a theorem that addresses this question (see Forster and Sober, 1994). The model selection literature explores this and other properties of AIC and other model selection criteria. In addition, Akaike's concept of predictive accuracy needs to be supplemented. Akaike described what Forster (2002) calls *interpolative* predictive accuracy; the concept of *extrapolative* predictive accuracy has interestingly different properties.

Notice that it doesn't matter to the Akaike framework or to AIC whether the models one considers are nested or non-nested. Comparing nested models makes sense because nested models can make different predictions when fitted to the data. It *does* seem strange to compare nested models if the goal is to discover which model is true. Since nested models are not in conflict, why does one have to choose? It is here that the Akaike framework is fundamental. Bayesians typically see truth as the goal of inference — the point of evaluating data is to say which of the competing theories one has formulated has the highest probability of being true. When predictive accuracy is substituted for truth as the goal of inference, the epistemological landscape undergoes a fundamental change.

7. Conclusion

Objective Bayesianism has its place and so does subjective Bayesianism. By 'objective Bayesianism' I don't mean a Bayesianism based on the Principle of Indifference (how could that be objective?), but one in which priors are objectively justified by a plausible account of a chance process. When I say that subjective Bayesianism has its place, I mean that agents who have degrees of belief in a proposition should use Bayes's Theorem to update. However, these two arenas for Bayesianism leave a large void in the theory of scientific inference. Many of the hypotheses of interest to science do not have objective prior probabilities. In addition, there are many composite hypotheses for which objective likelihoods cannot be provided. The reaction to these exigencies should not be a retreat to subjective Bayesianism. This is because it is doubtful that people always have subjective degrees of belief in hypotheses before they have any evidence in hand. But more importantly, the scientific enterprise aims to separate objective evidence from subjective preconception.

These problems come vividly into focus when they are brought to bear on the question of why simplicity matters in scientific inference. Bayesians have two resources to use in framing an answer. They can argue that simpler theories have higher prior probabilities or that they have higher likelihoods. When models are nested, it is impossible for the simpler model to have the higher prior (or posterior). For non-nested models, there is no logical contradiction in assigning simpler models higher priors, but what could justify that assignment? It is not enough that one *has* various prior degrees of belief. The question is why those assignments are right and others are wrong. Hopes for a likelihood account of the role of simplicity are likewise dim.[17] Models containing adjustable parameters are composite, and it often is obscure how the likelihoods of composite hypotheses can be compared objectively. One might be tempted to solve this problem by using the best-case strategy. However, this renders simpler hypotheses less likely, not more so.

Is it plausible to think that these problems for Bayesianism will be solved with more time and hard work? I tend to regard them as permanent and intractable. In contemplating the prospects for progress in this research programme, it is worth considering the fact that simplicity is not a puzzlement in the Akaike framework; rather, its justification is patent. There are no prior probabilities here, and the problem of evaluating the likelihoods of composite hypotheses does not arise. I don't want to suggest that this newer framework is a paradise free of conceptual puzzles. But it is a framework well worth exploring, in view of Bayesianism's scope and limits.

Note. I thank Martin Barrett, Ellery Eells, Branden Fitelson, Richard Royall, and Michael Steel for useful discussion.

References

Akaike, H. (1973), 'Information Theory as an Extension of the Maximum Likelihood Principle', in B. Petrov and F. Csaki (eds), *Second International Symposium on Information Theory*, Budapest: Akademiai Kiado, pp. 267–81.
Burnham, K. and Anderson, D. (1998), *Model Selection and Inference — a Practical Information-Theoretic Approach*, New York: Springer.
Crick, F. (1968), 'The Origin of the Genetic Code', *Journal of Molecular Biology* **38**: 367–79.
Edwards, A. (1972), *Likelihood*, Cambridge: Cambridge University Press.

---

[17] There is a special case in which I think this pessimism is misplaced. Cladistic parsimony is a method of inference used in phylogeny reconstruction. I suspect that this method makes sense to the extent that it reflects likelihood considerations; see Sober (1988, 2002b) for discussion.

Edwards, A. and Cavalli-Sforza, L. (1964), 'Reconstruction of Evolutionary Trees', in V. Heywood and J. McNeill (eds), *Phenetic and Phylogenetic Classification*, New York Systematics Association Publication No. 6, pp. 67–76.

Fisher, R. (1938), 'A Comment on H. Jeffreys's "Maximum Likelihood, Inverse Probability, and the Method of Moments"', *Annals of Eugenics* **8**: 146–51.

Forster, M. (1995), 'Bayes and Bust — Simplicity as a Problem for a Probabilist's Approach to Confirmation', *British Journal for the Philosophy of Science* **46**: 399–429.

Forster, M. (2000), 'Hard Problems in the Philosophy of Science — Idealisation and Commensurability', in R. Nola and H. Sankey (eds), *After Popper, Kuhn, and Feyerabend*, London: Kluwer, pp. 231–50.

Forster, M. (2002), 'In Defense of the Predictive Accuracy Framework', *PSA 2000 — Proceedings of the Philosophy of Science Association*.

Forster, M. and Sober, E. (1994), 'How to Tell when Simpler, More Unified, or Less *Ad Hoc* Theories will Provide More Accurate Predictions', *British Journal for the Philosophy of Science* **45**: 1–36.

Forster, M. and Sober, E. (2002), 'Why Likelihood?', in M. Taper and S. Lee (eds), *The Nature of Scientific Evidence*, Chicago: University of Chicago Press.

Hacking, I. (1968), *The Logic of Statistical Inference*, Cambridge: Cambridge University Press.

Howson, C. (1988), 'On the Consistency of Jeffreys's Simplicity Postulate and its Role in Bayesian Inference', *Philosophical Quarterly* **38**: 68–83.

Kalbfleisch, J. and Sprott, D. (1970), 'Application of Likelihood Methods to Models Involving Large Numbers of Parameters (with discussion)', *Journal of the Royal Statistical Society B* **32**: 175–208.

Lewis, P. (1998), 'Maximum Likelihood as an Alternative to Parsimony for Inferring Phylogeny Using Nucleotide Sequence Data', in D. Soltis, P. Soltis, and J. Doyle (eds), *Molecular Systematics of Plants II*, Boston: Kluwer, pp. 132–63.

McQuarrie, A. and Tsai, C. (1998), *Regression and Time Series Model Selection*, Singapore: World Scientific.

Popper, K. (1959), *Logic of Scientific Discovery*, London: Hutchinson.

Royall, R. (1997), *Statistical Evidence — a Likelihood Paradigm*, Boca Raton: Chapman and Hall.

Sakamoto, Y., Ishiguro, M., and Kitagawa, G. (1986), *Akaike Information Criterion Statistics*, New York: Springer.

Schwarz, G. (1978), 'Estimating the Dimension of a Model', *Annals of Statistics* **6**: 461–5.

Sober, E. (1988), *Reconstructing the Past — Parsimony, Evolution, and Inference*, Cambridge: MIT Press.

Sober, E. (1990), 'Let's Razor Ockham's Razor', in D. Knowles (ed.), *Explanation and Its Limits*, Royal Institute of Philosophy Supplementary Volume 27, Cambridge: Cambridge University Press, pp. 73–94.

Sober, E. (1998), 'Instrumentalism Revisited', *Critica* **31**: 3–38.

Sober, E. (1999), 'Modus Darwin', *Biology and Philosophy* **14**: 253–78.

Sober, E. (2002a), 'Instrumentalism, Parsimony, and the Akaike Framework', *PSA 2000 — Proceedings of the Philosophy of Science Association*.

Sober, E. (2002b), 'Reconstructing the Character States of Ancestors — A Likelihood Perspective on Cladistic Parsimony', *The Monist*.

3

Bayesianism in Statistics

COLIN HOWSON

I

1. The First Bayesian Theory

ONE PART OF THE REASON why the Bayesian Theory is eponymously titled is
that Bayes was the first to use it to solve a statistical problem. The other is
that Bayes is, though wrongly, credited with proving what is now known as
Bayes's Theorem. In fact, the credit for this must go to Laplace. What Bayes
did do, however, fell only an epsilon short of that, in giving something very
like a Dutch Book justification for the principle of conditional probability
$P(H|E)P(E) = P(H \& E)$, of which Bayes's Theorem is an immediate consequence (1763, Proposition 3). At any rate, having carefully presented and justified the basic principles of probabilistic reasoning, i.e. the usual probability
axioms, in an explicitly epistemic context, he used them to derive a posterior
probability distribution for a binomial parameter p.

Bayes's result solved an outstanding problem, that of 'inverting' James
Bernoulli's (Weak) Law of Large Numbers. Bernoulli's Theorem showed that,
subject to the so-called iid conditions (independence and identical distribution
from trial to trial), the probability that the difference between a fixed probability p of 'success' and the sample average s is less than some pre-assigned
magnitude tends to 1 as n, the number of trials, tends to infinity. For many people, including Bernoulli himself, who had been looking for a way of determining probabilities a posteriori, this did not quite do the job: in today's
jargon, Bernoulli made p the parameter and s the random variable whose probability distribution depends on p and n. Bayes in his famous 1763 *Memoir*
seemed to have obtained the desired inversion of Bernoulli's Theorem, by giving a probability distribution for p as a function of s and n.

To obtain his result Bayes found that he had to use a uniform, or constant,
prior probability distribution over the range of values p, i.e. the closed unit

Proceedings of the British Academy, **113**, 39–69. © The British Academy, 2002.

interval. It seems that Bayes himself was not entirely happy about the justification for this; his discussion occupies a Scholium, in which he appears to advance something like the principle that if your background information is neutral between the possible values of p, then that neutrality should take the form of a uniform prior density. Bayes's own appearance of tentativeness in adopting this principle,[1] later called by Keynes the Principle of Indifference, seemed subsequently vindicated by a disturbing consequence of the Principle, albeit one relatively slow in coming to light. When it did, however, it seemed increasingly to cast doubt on the entire Bayesian methodology. The consequence is this. If you know nothing about p then you presumably know nothing about $f(p)$, where f is any continuous invertible tranformation of p, e.g. p^2. But p and p^2 cannot both be uniformly distributed. It seems to follow that no probabilistic notion of complete epistemic neutrality can be well defined. Not only that: p's possible values conveniently form a bounded interval. But there are other statistical parameters of equal if not more interest which do not. But for any parameter capable of taking any real-number value, or even any positive value, no uniform distribution is possible which obeys the laws of probability. Yet Bayes's Theorem tells us that some prior distribution must be assumed if a posterior probability is to exist.

These consequences were slow to become appreciated, however, and after Laplace had rediscovered Bayes's result and adopted the Principle of Indifference with none of Bayes's misgivings (he used it famously to *define* probability as the ratio of the number of favourable to the number of all possible outcomes), the theory of posterior distributions based wherever possible on uniform prior distributions became the dominant methodology for inductive inference. Nevertheless, by the early years of the twentieth century the objectionable features alluded to, particularly the vulnerability of the Principle of Indifference to transformational problems, were regarded as sufficiently disturbing to suggest that something was fundamentally wrong. But it is a commonplace that no theory is rejected until a better one appears. And it duly did. In the early years of the twentieth century another approach to epistemic probability was in fact germinating, not appealing in any way to the Principle of Indifference. It was not this theory that succeeded the First Bayesian Theory, however. Even had it been more widely known it would no doubt have seemed, as to many it still seems, to throw the baby of determinate answers to practical questions out with the bathwater of the Principle of Indifference, for in offering nothing to replace the Principle it amounted to no

---

[1] If indeed that was the principle he adopted. Commentators are divided on this point. See Earman (1992), Chapter 1.

more than the probability calculus itself interpreted as a theory of consistently distributed degrees of belief.

2. R. A. Fisher and significance tests

I shall describe this Second Bayesian Theory, or at any rate the version of it which in my opinion best justifies its normative status (actually as an authentic logic), in some detail in the second part of this paper. For now it is sufficient to note that however likely or unlikely its acceptance at the time, it was anyway pre-empted by a series of brilliant papers by a young statistician and geneticist, R. A. Fisher. At the same time as drawing further attention to the defects, which he regarded as insuperable, of the First Bayesian Theory, Fisher advanced a radically different, non-probabilistic, methodology for testing and estimation in statistics. His ideas were summarized and broadcast to a wider audience in two books, *Statistical Methods for Research Workers* (1926) and *The Design of Experiments* (1935), the former of which became for years the handbook of applied statistics. The latter contains the description of a thought-experiment that became justly famous, presenting in a simple and easily understood way all the basic ideas of Fisher's new theory of significance tests, tests which remain to this day the most widely used tool for testing statistical hypotheses.

In Fisher's experiment a lady who claims to be able to discern by taste whether a cup of tea has had the milk or the tea added first is asked to distinguish the milk-first from the tea-first in a randomized sequence of eight cups, four of which have the milk added first and the remaining four the tea. The null hypothesis is that she has no power of discrimination in the way she claims, and the randomization of the order in which the tea-first and the milk-first cups are presented is supposed to ensure that if the null hypothesis is true, then any success is due to chance, and a chance moreover which can be precisely calculated. Most of the ingredients of a classic (one-tailed) test of significance are present in this example: an imprecise background hypothesis that some effect may be due to a systematic cause, in this case the power of discrimination by taste; there is the null hypothesis which is the negation of that hypothesis; there is an experimental design, usually involving randomization to ensure a definite chance distribution over a suitable outcome space (in this case the number of correctly identified milk-first cups) if the null hypothesis is true.

What remains is to compute the probability of obtaining an outcome at least as discrepant with the null hypothesis (where the direction of discrepancy is supplied by the alternative) as that actually observed if the null hypothesis is true. If this probability, now usually called the *P*-value associated with that outcome, is less than some appropriately small figure (often 5 per cent, determining approximately two standard deviations either side of the mean if the chance distribution is normal), then the null hypothesis is rejected at that level of significance.[2] The force of the rejection, according to Fisher, is 'logically that of the simple disjunction: *Either* an exceptionally rare chance has occurred, *or* the theory of random distribution [i.e. the null hypothesis] is not true' (1956, p. 39). Granted the implicit premise that such unlikely events can be safely regarded as not occurring in the context of *this* particular experiment, the rejection is sound and the null hypothesis can be regarded as 'definitely disproved' by the significant outcome (1932, p. 83).

This procedure, though at first sight plausible, does however raise some questions. Why, for example, is the 'rare chance' not that of the observation itself, but of the observation *together with all those possible values at least as extreme*? And what is the systematic relation between the level of significance and the confidence we are entitled to in rejecting at that level? The former is the chance, given the null hypothesis, of observing that outcome or a worse one; the latter would seem to be just the reverse: the chance, given the observed outcome, of the null hypothesis being true. A partial answer to the first question is that where the probability distribution is continuous its density distribution can be changed to any form, including a uniform one, by an appropriate transformation of the variate, so one cannot appeal to the magnitude of the ordinate of the density curve. Non-degenerate regions, however, have *probabilities*, which are invariant. But this still leaves the question of why the region deemed significant should be the *tail* region: it would seem likely that invoking the smaller probabilities, or densities, of all the outcomes *worse* than the one actually observed will unfairly weight the verdict drawn from the observation against the null hypothesis. So indeed it will turn out, as we shall see.

An answer to the second question, of the relation between level of significance and level of confidence, was given by Fisher himself: it is that if we reject the null hypothesis at a small significance level, say 5 per cent, then in the long run we should make a mistaken rejection only once in every twenty

---

[2] Only a single outcome in the tea-tasting example, namely getting all four identifications correct, would be significant at the 5 per cent level (the probability distribution given by the null hypothesis over the five possible outcomes is symmetrical: if X is the number of correct identifications then $X=0$ and $X=4$ occur with probability 1/70; $X=1$ and $X=3$ have probability 16/70, and $X=2$ has probability 35/70; $X=0$ is not significant as it does not discriminate between the null and the background hypothesis).

times on the average (1932, p. 45). This answer, often called the 'repeated-sampling' justification, and which anticipates the later ideas of Neyman and Pearson, raises the further question of the relation between the frequency of mistakes and the confidence you should feel in rejecting the null hypothesis. The possibly surprising answer is that these are *independent* of each other: the proportion of times you would be wrong in rejecting the null hypothesis on data *D* is completely independent of the size of the chance, given *D*, of the null hypothesis being true. We shall soon see a striking instance of this.

Note the emphasis on *falsification* in Fisher's account. This is not accidental. Fisher was, at roughly the same time as Popper, also advancing the case for what philosophers have come to call falsificationism, though unfairly Fisher has never been credited by philosophers with co-authorship of the idea. Both these thinkers believed that testing could show only whether observational data are in agreement with the hypothesis tested: sufficiently well-specified hypotheses may be refuted by observation in suitably designed experiments, but never rendered probable by them. If the hypothesis survives the test, that means simply that it stands as unfalsified until further tests can be devised (Fisher, 1932, p. 9).

Like Popper also, Fisher offered a positive theory of inference to balance the negative thrust of his testing theory. Fisher's had three rather different parts. One was his theory of likelihood as a measure of rational confidence, whose formal dissimilarity to probability he stressed, just as did Popper with respect to his own likelihood-based theory of corroboration. Second, there was a theory of estimation. A significance test may tell you when a null hypothesis about the value of a parameter, for example, is to be rejected, but it does not give any precise information about the true value. The theory of estimation was intended to remedy the deficit, advancing criteria for 'good' estimators including minimum variance (efficiency), retention of all the sample information about the parameter (sufficiency), convergence in probability to the true value (consistency) and in particular maximum likelihood; these are of course not all independent of each other. To anyone not swept away by the mathematical virtuosity of Fisher, the very idea of making a point estimate, as opposed to an interval one, on the basis of any finite set of observations might be questioned on the common-sense ground that it is exceedingly unlikely to be correct, and indeed the temptation is to say that with probability 1 it will not be. An interval estimate would sound more reasonable, and the third strand in Fisher's positive theory is nothing less than a theory of a posteriori probabilities for suitable intervals. The striking feature of *fiducial probability*, as he called it, is that it does not depend in any way on probabilities a priori, i.e. on the apparatus of Bayesian probability whose use Fisher condemned as, if not inconsistent, then 'measuring merely psychological

tendencies, theorems respecting which are useless for scientific purposes'
(1935, pp. 6, 7). But confusing as it seemed to the role of scalar parameter and
random variable, fiducial probability was dismissed by many as fundamentally mistaken, and never gained widespread acceptance.

Despite the questionable aspects of Fisher's theory I have drawn attention
to, the Fisherian revolution was remarkably successful, its central features
persisting to this day as the foundation of testing and estimation. Yet his ideas,
combined with the idea of power and the theory of confidence intervals added
by Neyman and Pearson to form the amalgam sometimes called frequentism,
sometimes classical statistics, have come under increasingly severe attack,
and its principal antagonist is none other than the method of inverse probability so powerfully discredited by Fisher. This surprising reversal of history
is due to two principal causes. First, it became clear that a great deal of this
Second Bayesian Theory, indeed the part that today is regarded as its core,
was not directly hit by the problems associated with the Principle of
Indifference, and that its explicit incorporation of the scientist's own evaluation of the prior data is not necessarily a bad thing. Second, just as Fisher and
others before him had derived paradoxical consequences from the First
Bayesian Theory, so consequences no less paradoxical were shown to follow
from Fisher's own theory, to which the Second Theory is demonstrably
immune. Thus, (1) a classical (Fisherian or Neyman–Pearson-type) statistician can plan, with probability 1, to obtain an outcome significant at any
assigned level if they are prepared to continue sampling up to large enough,
but with probability 1 *finite*, values of n. By contrast, assuming that no
improper priors are used, and that the probability function is countably addi-
tive, no Bayesian (I shall henceforward use the unqualified adjective
'Bayesian' as it is mainly used today, to refer to the Second Bayesian Theory
or to a subscriber to it) can plan, with probability 1 of success, to achieve any
prespecified posterior probability (see Kadane et al., 1996). (2) *One and the
same observation* may and may not be significant depending on what is taken
to be the outcome space of the experiment (Howson and Urbach, 1993,
pp. 182–3). This cannot happen in the Bayesian Theory since only the test
outcome itself is relevant to determining the posterior probability (this is
essentially just the Likelihood Principle[3]).

Not only is the Bayesian Theory immune to the problems which beset the
classical approach: weak though it might appear in offering only a standard
of consistency for probability assignments, it is strong enough to make

---

[3] For a statement of this principle, and a proof of it from two plausible but by no means specifically
Bayesian assumptions about evidential support, see Birnbaum (1962).

damaging comparisons with the classical alternative, in particular being able to diagnose which feature of that theory is responsible and why. We shall see this in the discussion which follows of possibly the most striking of all the paradoxical consequences of the classical theory, discovered by Dennis Lindley over forty years ago: (3) *It is possible to find an outcome significant at any degree of significance, however small, which would make the null hypothesis overwhelmingly probable, according to almost any way of assigning prior probabilities.* Though Lindley's result has been much discussed by statisticians, and the subject of at least one full-scale conference, it still merits only a passing reference in most Bayesian texts, and often none at all in non-Bayesian ones. I myself believe that one day it will be credited as one of the decisive contributions to statistical inference. For this reason, and because it provides a background to other topics arising in this paper, I shall discuss it in some detail.

3. Lindley's Paradox

There are actually two main theories of significance testing: Fisher's based on *P*-values already described in the illustrative context of the tea-tasting lady; and the Neyman–Pearson theory of error probabilities. Lindley's discussion begins against the background of a typical Fisherian test. The scenario is a known model, in which the data are a sample from a specified distribution with one unknown real-valued parameter θ. The test is a test of the null hypothesis H_0 that θ takes the value θ_0. A sequence of observations $X = (x_1, \ldots x_n)$ is determined from which a test statistic $T(X)$ is constructed.[4] $T(X)$ is supposed to carry all the information in X about θ, and large values of T are supposed to indicate increasing 'discrepancy' with H_0.[5] Suppose the model is a normal distribution with 'known' variance σ^2 and θ is the mean. If there is no 'directional' vague alternative to the null (as in the tea-tasting where the conflict with the null is focused on the right tail), T is usually taken to be the absolute value of the difference between the sample mean and θ_0 measured in units of the standard error, i.e. $T(X) = |\overline{X} - \theta_0| \sqrt{n}/\sigma$, where $\overline{X}$ is the mean of X. The *P*-value of a value t of T is the 'tail-area' probability $P(T \geq t | H_0) = 2(1 - \Phi(t))$, where Φ is the standard normal distribution function. Let λ_α be the value of

---

[4] I shall follow the common practice of using upper case to signify the random variable and lower case its values.

[5] Larger values of T are supposed to reflect greater 'distance' from what is predicted by the null hypothesis, where some non-specific alternative may, as in the tea-tasting example, give a preferred direction to this distance.

t such that $P(T \geq t|H_0) = \alpha$. Two observations are important for the subsequent discussion: (a) λ_α does not depend on *n*, and hence (b) as *n* takes larger values (though in any given experiment *n* is of course fixed) the same observation $T = \lambda_\alpha$ corresponds to values of $\overline{X}$ increasingly close to θ_0.

We already know what a *P*-value is supposed to signify: the extent to which the observed discrepancy *contradicts* H_0. Fisher himself used this terminology repeatedly, and also made the following statement: 'Tests of significance do not generally lead to any probability statements about the real world, but to a rational and well-defined measure of reluctance to the acceptance of the hypotheses they test' (1956, p. 44). Lindley's result casts severe doubt on these claims, by showing how an observation significant at say the 1 per cent level comes eventually to be very strong evidence not against but *for* the null hypothesis. In the simple example he used, H_0 is the hypothesis that the mean θ of a normal distribution of known variance σ^2 takes a particular value θ_0. Where the alternative is simply $\theta \neq \theta_0$ the statistic used is *T* above; where it is $\theta > \theta_0$ it is *T* without the modulus. Either can be assumed here. Now suppose that the observed value *t* of *T* is λ_α, i.e. the value of *t* such that $P(T \geq t|H_0) = \alpha$. Suppose, finally, that a positive prior probability *c*, of any magnitude, is placed on H_0, with the remainder $1 - c$ spread uniformly over an interval *I* containing $\overline{\theta}$ (the assumption of uniformity over a bounded interval is for computational simplicity only; the result does not depend on it). $\overline{X}$ is a sufficient statistic for θ and hence the posterior probability $\overline{c}$ of H_0 given any value $\overline{X}$ is the same as if it is given *X*. By Bayes's Theorem

$$\overline{c} = \frac{c \exp[-(1/2)\lambda_\alpha^2]}{c \exp[-(1/2)\lambda_\alpha^2] + (1-c)m^{-1} \int_I \exp[-(n/2\sigma^2)(\overline{x}-\theta)^2]d\theta}$$

where *m* is the length of *I*. For sufficiently large *n*, $\overline{x}$ will be well within *I*, since $\overline{x} - \theta_0$ tends to 0 (see (b) in the preceding paragraph), and so the integral is approximately $\sigma\sqrt{(2\pi/n)}$. Since λ_α does not depend on *n*, $\overline{c}$ tends to 1 as *n* increases. In other words, for any α and $\overline{c}$ between 0 and 1 there is an *n* and an outcome $\overline{x}$ such that $\overline{x}$ is significant at level ε and the posterior probability of H_0 on $\overline{x}$ exceeds $1 - \varepsilon$. As Lindley remarks, 'The usual interpretation of [the significance level] is that there is good reason to believe $\theta \neq \theta_0$; and of [the posterior probability] that there is good reason to believe $\theta = \theta_0$' (1957, p. 187).

This is Lindley's Paradox. It seems to be a convincing refutation of any claim (recall the discussion in the previous section) that because a wrong decision would be made with a sufficiently small frequency, that frequency can be taken as an index of one's warranted confidence in the rightness of the decision. This obviously applies equally to the classical theory of confidence intervals, since the $1-\alpha$ confidence interval obtained from $\bar{x}$ excludes θ_0, which lies on a boundary, and so has posterior probability less than $1-\bar{c}$ of containing the true value of θ.

We can look at the result in another way that spells out what rejection by a large sample actually amounts to. If we equate α to approximately the posterior probability $P(H_0|\bar{x})$, as the standard interpretation of P-values itself suggests, then we find that we must have $c = (1+k\sqrt{n})^{-1}$ for some k depending on α. This implies that significance tests incorporate an increasing bias against the null hypothesis as the sample size increases, to the point where, for large enough n, α can consistently indicate our degree of confidence only if our prior probability for H_0 is practically 0. This means that practically all the 'evidence' against the null hypothesis from a significant result is actually supplied by the prior probability! Thus, contrary to the received view of reliability increasing with sample size, 'significant' outcomes actually provide *weaker* evidence against the null hypothesis, other things being equal, as the sample size increases, to the point where they start to provide evidence in its favour.

It is not of course to be expected that significance tests will always give misleading results; they would hardly have become so entrenched if they did. In fact, as Lindley shows, in a range of typical cases they are in broad agreement with the posterior probabilities. Following Lindley, let

$$A = cm\exp[-(1/2)\lambda_\alpha^2]/(1-c)\sqrt{(2\pi)}.$$

Then $\bar{c} = A/(A+\sigma/\sqrt{n})$ and so gets small as n gets small (though the approximation is not valid for very small n). At some point, therefore, $\bar{c}$ must agree with the significance level, and the agreement can be fairly stable over a wide range of sample sizes (see Lindley's paper for examples).[6]

It might be objected that these conclusions depend on being prepared, as Fisher was of course not, to admit a posteriori probabilities as meaningful. This is not a valid objection. First, there now exists a demonstrably consistent theory of a posteriori probability, and a fortiori a meaningful one. Second, Lindley's result holds for all prior density distributions which do not take extreme values (0 or infinity) in the neighbourhood of $\bar{x}$. In other words, it makes little or no prejudicial assumption in the prior distribution over the alternative to H_0 which, given the underlying normal model, we can write as the negation $-H_0$ of H_0.

---

[6] Even so, the overestimation by the significance level of the evidence against the null hypothesis relative to the posterior probability on entire classes of 'reasonable' prior distribution can be very marked even for quite moderate samples. Phillips has some nice examples (1973, pp. 347–50), and Berger and his co-workers have produced some very general results on these lines (Berger and Sellke, 1987 and Berger and Delampady, 1987).

Where there might be cause for concern, however, is in placing a lump of prior probability on H_0. Indeed, a natural line of defence against Lindley's Paradox, and one which does not simply rule Bayesian probability out of court, is the frequent claim that significance testing is not really testing whether θ is *exactly* equal to θ_0, which as a matter of fact nobody really believes, but merely whether θ is in a small neighbourhood of θ_0. Thus the alleged disagreement between significance and posterior probability for large samples is based on a false premise, for no statistician would, in Bayesian language, put a positive prior probability on θ_0. What is really being tested is a composite null. And it turns out that for composite nulls with a continuous prior distribution there is no necessary disagreement at all (Berger and Delampady, 1987, p. 322).

There are three answers to this. (a) It is not true that nobody ever gives credence to a sharp null hypothesis. On the contrary, it is easy to think of cases where they would be given a substantial degree of credibility, for example where causal links of the sort predicted by the main alternative theory are thought highly implausible (e.g. telepathy), or where they are predicted by a body of theory as the mean value of an error distribution. (b) The objection misses the point, which is that, to judge from their own words, Fisher and many other practitioners regarded and continue to regard significance testing as sound independently of how credible the null hypothesis is thought to be; for example Kendall and Stuart (1973, p. 190) simply describe the test as 'valid'. (c) To deny that the null hypothesis has a non-zero chance of being true is to make a very strong statement indeed, implying that *no possible evidence* could ever make you change your mind. How many disbelievers in the point null would be willing to accept that level of a priori dogmatism?

Far from being prejudicial, Lindley's example shows, I think, how the apparatus of Bayesian probability is an indispensable explanatory tool: it tells us in an intuitively appealing way not only that something is wrong with significance tests but also what and why. To that I now turn. As we saw earlier, the basic inferential principle of significance tests is that events which are extremely rare according to H should be regarded as evidence against H should they occur. According to Fisher, as we saw, the force of the rejection is the disjunction: either an extremely rare event has occurred or the null hypothesis is false, with the tacit premise that such events can be regarded as practical impossibilities. As we also noted in the earlier discussion, there is a curious feature to Fisher's own use of this idea, namely his characterizing the outcome as a *disjunctive tail-event*; but there are also other features no less problematic. Let us look at them in turn.

(1) Why should the 'extremely rare event' be characterized not as the observed outcome itself but as a disjunctive class of which that outcome is merely one

member? Admittedly, the chance of $\bar{x}$ itself cannot be greater than the probability of the tail region determined by $\bar{x}$; indeed, for continuous distributions it is zero, as are all the other individual outcomes. But while the P-value of an outome significant at level α is of course α whatever the size of the sample, the density ordinate given H_0, i.e. the likelihood, is $\sqrt{(n/2\pi\sigma^2)}\exp[-(1/2)\lambda_\alpha^2]$ which increases and tends to infinity with n. This means that outcomes in the neighbourhod of $\bar{x}$ are increasing in probability relative to the more extreme ones. Intuitively, that is surely the relevant 'chance'. It is, but not by itself, and this leads on to another problematic feature of Fisher's theory.

(2) This attaches to Fisher's inference rule itself: it fails to take into account how improbable the outcome is under alternative hypotheses. Yet such a consideration is always highly relevant. The reason it got ignored is probably that Fisher knew that the alternatives to the null are infinite in number, and also knew that no non-Bayesian theory could regard the probability $P(E \mid -H_0)$ as meaningful, since it is an average of likelihoods weighted by prior probabilities. So it got left out of the picture: only the chances determined by the null hypothesis are relevant to its evaluation. But the resulting impoverishment impugns the soundness of the rule: if the probability, or probability density, of the observed data E given that H_0 is false is smaller than that on H_0, then E does not tell against H_0 at all: on the contrary, it *supports* H_0. When E is properly described (in terms of T) as λ_α and not as a tail region we see that $L(E \mid H_0)/L(E \mid -H_0)$, the likelihood ratio known as the Bayes factor, is equal to $m\sqrt{(n/2\pi\sigma^2)}\exp[-(1/2)\lambda_\alpha^2]$, which goes to infinity with $\sqrt{n}$.[7] In other words, though the outcome when correctly described is unlikely given H_0, it is so much less likely on a weighted average of the alternatives to H_0 that in the limit the probability of H_0 goes to 1.

To sum up: it is by a combination of misdescription of the outcome and ignoring its chance on the alternatives to H_0 that the significance test increasingly delivers the wrong verdict as sample size increases; and this is what Lindley's Paradox shows so dramatically.

4. Neyman–Pearson

The Neyman–Pearson theory, unlike Fisher's, does explicitly compare the null hypothesis with alternatives, and employs a quite different rationale for

---

[7] In view of the fact that Jeffreys had noticed that the Bayes factor increases with $\sqrt{n}$ for a constant deviation in units of the standard error (1961, p. 248), Lindley's Paradox is sometimes attributed to Jeffreys. I think this slightly unfair, since Jeffreys did not seem to notice, or was not interested in pointing out, that this implied an asymptotic conflict with Fisherian significance tests.

significance tests. These are now decision rules represented by a rejection region R in the sample space, and their utility is measured in terms of their characteristic frequency of errors of the first and second kind (of rejecting a true hypothesis and accepting a false one respectively). Where there are just two competing hypotheses H_0 and H_1, that $\theta = \theta_0$ and $\theta = \theta_1$ respectively, the type 2 error is to accept H_0 when H_1 is true. For a fixed experimental size these frequencies cannot be controlled simultaneously, and for this reason R is determined by minimizing the chance β of a type 2 error relative to a pre-selected magnitude α, the significance level, of the chance of a type 1 error. According to the Fundamental Lemma of Neyman and Pearson, the resulting R is determined by the likelihood ratio $L(\bar{x}|H_0)/L(\bar{x}|H_1)$ being less than some quantity r depending on α, which in view of the above diagnosis of Lindley's Paradox sounds as if the Neyman–Pearson theory is on the right track. But the inequality determines a right-tail region (for the example given) when $\theta_1 > \theta_0$, and in describing the significant outcome in terms of belonging to a (rejection) *region* it still falls prey to Lindley's Paradox. Consider the simplest case where there is one 'simple' alternative, H_1, in the frame of seriously considered alternatives, and H_1 is the hypothesis that $\theta = \theta_1$ where $\theta_1 > \theta_0$. Suppose that the size α of the rejection region C is kept constant and n allowed to vary. Suppose as before that $\bar{x}$ is observed on the boundary of C. The power of the test tends to 1 as n increases, and as before it rejects H_0. Suppose the prior probabilities of H_0 and H_1 are positive. The Bayes factor is $p(\bar{x}|H_0)/p(\bar{x}|H_1) = \exp[\Delta\sqrt{n}(\Delta\sqrt{n} - 2\lambda_\alpha)/2]$, where the 'effect size' $\Delta = (\theta_1 - \theta_0)/\sigma$, which clearly tends to infinity, making the posterior odds in favour of H_0 infinite.[8]

The usual response is that Lindley's Paradox depends on keeping α constant between experiments, for which there is no justification in the Neyman–Pearson theory where the emphasis is merely on tests which make α and β (the chance of a type 2 error of accepting H_0 when H_1 is true) as small as possible. When n is fixed it is well known that α and β cannot be controlled simultaneously, but by increasing n and holding the likelihood ratio r constant, α and β can both simultaneously be reduced. Indeed, this is the only way in which both chances of error can be simultaneously reduced (Barnard, 1990, pp. 611–12; the observation is originally due to Pitman). This fact is illustrated by the example above (the discussion is taken from Freeman, 1993, pp. 1447–8). The rejection region for H_0 against H_1 is an interval $I(r) = [(\theta_0 + \theta_1)/2 - \sigma\log r/(n\Delta), \infty]$; in practice r is fixed by the condition

---

[8] Hence for data just significant at the $100 \times \alpha$ per cent level, if θ_0 is very close to θ_1, then for moderate σ a correspondingly large sample will be needed to discriminate significantly in terms of posterior odds between H_0 and H_1.

that $P(I(r)|H_0) = \alpha$ for some suitably small α. From this equation we can deduce the following explicit definition of α in terms of n and r:

$$\alpha = \Phi(-\Delta\sqrt{n}/2 + \log r/\Delta\sqrt{n}),$$

Similarly,

$$\beta = \Phi(-\Delta\sqrt{n}/2 - \log r/\Delta\sqrt{n}).$$

If r is chosen to give $\alpha=5$ per cent when $n=10$, then with r held constant and $\Delta = 0.5$, both α and β strictly reduce to 0 as n increases beyond 5; smaller values of Δ merely mean larger threshold n. Now it seems as if Lindley's Paradox is avoided, because by Bayes's Theorem the value $\bar{x}(r)$ on the boundary determined by constant r determines exactly the same posterior odds ratio for H_0 against H_1 in all larger experiments, and the posterior probability of H_0 does not tend to 1.

That may be true, but it only serves to undermine any connection between error probabilities and the inferences which should be drawn from observations in tests devised according to them. For the claim that keeping the likelihood ratio determining the rejection boundary constant and increasing n makes for an epistemically more informative test, because it will have smaller α and β, runs up against the fact that by Bayes's Theorem an observation $\bar{x}(r)$ on the boundary determined by constant r must give the *same* posterior odds ratio for H_0 against H_1 whatever the size of the sample, i.e. *however small α and β might be*. Hence as far as the laws of probability are concerned there is simply no warrant for claiming that observing $\bar{x}(r)$ in a bigger test with smaller α and β is in any way more informative than in a smaller test. Once again we see how appeal to tail-region probabilities is systematically misleading, whether they are error probabilities or not. In fact, the posterior probability can easily be shown to be independent of the error probabilities. A simple illustration is the famous Harvard Medical School Test,[9] where a rejecting outcome (the patient tests positive) can be consistent with an arbitrarily large posterior probability of the null hypothesis (that the patient does not have the disease), however small the error probabilities; here of course one has to fix the prior probability appropriately.

5. Likelihood

Bayes's Theorem tells us that it is not membership of any region which determines how the outcome discriminates between the competing hypotheses,

[9] Casscells et al. (1978).

only the value of the corresponding *likelihood ratio*. Given that, it might be asked why the Bayesian Theory is any better than a theory of evidence (for example Edwards's 1970) based *only* on likelihood ratios. First, because likelihood ratios by themselves do not tell you anything. It is true that Edwards and others have defined them, or more precisely their logarithms, as measures of comparative support. But this is merely a definition, true by fiat. Furthermore, it is easy to produce examples where the use of likelihood ratios as measures of support would give quite the wrong information. A completely bizarre hypothesis manufactured to 'predict' the observation would be best supported of all, quite contrary to the truth. The hypothesis that God willed a particular sequence of heads and tails in a long sequence of random-looking tosses of a fair-looking coin is not regarded by most people as supported by the evidence compared with the hypothesis that the sequence is a sample from $B(n,p)$ for appropriate p (Good, 1983, p. 132 has another nice example; and see the quotation there from de Finetti). *Assessments of support depend on relevant prior information.*

Edwards actually acknowledges this and his theory incorporates a notion of prior support satisfying the relation

posterior support = prior support + experimental support

an equation he acknowledges looks very much like the Bayesian

log-posterior odds = log-prior odds + log-likelihood ratio,

the more so since his supports are all logarithms (1970, p. 36). But this is effectively selling the pass. True, Edwards requires the prior odds to be 'objectified' as those which you would have made on the basis of an unperformed experiment, but even where this could be done, (a) it is still a subjective matter of how you believe such an experiment would turn out, and (b) it invokes entirely imaginary situations, where the Bayesian Theory does not.

The second reason why the Bayesian Theory is superior to a likelihood-only account is that there is no non-arbitrary sense to be made of the likelihood of a compound hypothesis, yet at the same time likelihood ratios *require* the full apparatus of prior probabilities where the alternative to the principal hypothesis under consideration is composite, as in practice it is more frequently than not. It is, for example, in Fisher's tea-tasting test, and for that matter in Lindley's own example, where there is one sharply defined hypothesis of interest, call it H_0 again, and a non-specific alternative, in this case $-H_0$. This sort of scenario is sufficiently representative for it to be used in an exemplary way by Fisher, and indeed it occurs time and again in practice. If it is conceded that likelihood ratios, in other words the comparison of likelihoods on a common scale, are important at all, we immediately have the problem of evaluating the denominator $P(\bar{x}|-H_0)$ of the likelihood ratio $P(\bar{x}|H_0)/P(\bar{x}|-H_0)$. To do this in any way that is not the merest hand-waving means evaluating an integral $\int f(\bar{x}|\theta)g(\theta)d\theta$ over the range of values of θ in the complement of H_0, where $g(\theta)$ is a prior density. Conceding the central importance of the likelihood ratio thus *inevitably* leads you to the Bayesian Theory, since it is the only theory in which probabilities of the form $P(\bar{x}|-H_0)$ are meaningful.

6. Priors

But it does of course also raise the question of the status of these prior densities, and not only of densities *within* a given model but also of the prior probability of the model itself. How are all these priors to be determined? On the consistency view of the probability axioms the question is easily, indeed obviously, answered: the appropriate prior distributions are simply those that reflect your beliefs about the subject matter, conditioned as these will presumably be by your available background evidence. But this answer has always seemed problematic, even to those Bayesians who would otherwise accept that view of the theory. The problem is that it seems to put everyone's views on a par, whether they are guided by the 'best' scientific theories and the evidence for them, or merely whimsical. It seems to lead to just that subjectivism that Fisher ruled out of bounds to science. It has been argued (most persistently by I. J. Good) that allowing explicitly for a subjective input is actually a positive aspect, since it is inevitably there and all the allegedly objectivistic theories of inference do is sweep it under the carpet (this is the methodology of SUTC as Good puts it in Good, 1983, p. 23). One could take this point further by adding that a proliferation of different viewpoints is usually valuable, if not indispensable, for progress: for example, if everyone had taken Einstein's view of quantum mechanics, it would probably not have developed as fast and as far as it has done. But maybe we wouldn't want to take L. Ron Hubbard's views (if he had any on quantum mechanics) equally seriously. And therein lies, or seems to lie, the problem.

On the other hand, attempts to produce criteria for determining 'objective' prior distributions have also been problematic. There is an enormous literature on the subject, and I shall just comment briefly on the two criteria which seem most to have commended themselves: 'informationlessness' and simplicity.

6.1 'Informationless' priors

The idea here is to use priors which 'let the data speak for themselves'. Uniform priors are an extreme example in so far as they are alleged to treat all possibilities as a priori equal. As we noted earlier, such priors fall foul of transformational problems when the possible cases form a continuum. And not only then. There is a presupposition that 'the class of all possibilities' is something that can be given uniquely, and it does not require much reflection to see that this is almost never the case, even when discreteness is assumed. Here is a simple example. Consider the possible cases concerning the distribution of a property B among three individuals. One way of determining these is as a fourfold partition: none has *B*, one has *B*, two have *B*, and three have *B*. On the other hand, if we give distinguishing names to the individuals, e.g., *a*, *b* and *c*, then the number of possible cases multiplies: the previous category 'one has *B*' divides into three, as either *a* has *B* or *b* has *B* or *c* has *B*, while the category 'two have *B*' divides into six. Neither way of describing these cases is more correct a priori than the other, and it is not difficult to reproduce the same sorts of contradictory results usually associated with continuous applications of the Principle of Indifference here too (as in Howson, 2001).

A less obviously problematic approach to informationlessness is to use the weaker condition merely that the prior distribution be dominated by the likelihood where it is appreciable. Such priors are also called *reference priors*. Uniform distributions over large enough intervals, where they can be consistently defined, generally have this property. Sometimes, however, a uniform distribution over the entire real line (for a location parameter), or a log-uniform distribution (for a scale parameter) over the half-line, are adopted. These are *improper* (i.e. their integrals diverge), and hence inconsistent with the probability axioms, but their employment is usually condoned on the ground that they can often be combined with the likelihood to yield a proper posterior distribution, and then they can be thought of as representing the *local* behaviour of a proper reference prior.

In addition, reference priors may be instances of an invariant form like the square root of the Fisher information

$$I(\theta|X) = -E\partial^2 \log p(x|\theta)/\partial \theta^2$$

Given that the right-hand side is equal to $E(\partial \log p(x|\theta)/\partial \theta)^2$, it is not difficult to see that if the prior density $p(\theta)$ is $\sqrt{I(\theta|X)}$ and $\phi(\theta)$ is any differentiable function of θ, then $p(\phi) = \sqrt{I(\phi|X)}$. Where the likelihood is normal, the invariant priors for the mean and standard deviation are both improper: the former is the uniform distribution and the latter the log-uniform. There are well-known problems with appealing to invariance, despite its attraction in the light of the problems plaguing the Principle of Indifference: there is not a unique invariant form, for one thing (Jeffreys, 1961, pp. 179–92), and any particular choice will be fairly *ad hoc*. Second, it may not always be well defined; the one above is not if the expectation does not exist. Third, the Jeffreys prior does not satisfy the Likelihood Principle, since it depends on the entire range of X. It also has some counterintuitive consequences, e.g. the joint prior for the mean and standard deviation of a normal likelihood is not the product of each separately; in other words, they are not independent, which intuitively they should be.

Practical problems are one thing, but there is a much deeper, in principle, problem with this whole approach of 'letting the data speak for themselves'. Doing that *is impossible*. To use an example of Jeffreys, Galileo's law says that a body in free fall from rest travels a distance $s(t) = (1/2)gt^2$ after t seconds. Suppose this has been verified for elapsed times $t_1, \ldots, t_n$. The data obviously also satisfy the bogus free-fall law $w(t) = (1/2)gt^2 + h(t)(t - t_1) \ldots (t - t_n)$, where $h(t)$ is any function of t not identically 0 (Jeffreys, 1961, p. 3). Which of these relationships do the data 'say' it is a sample of? The correct answer is of course that evidence by itself does not 'say' anything at all about the likely truth values of rival explanatory hypotheses. The posterior distribution is obtained from the data only via a prior distribution, and depending on the form this takes the posterior distribution will vary. Thus, for example, a uniform prior is just as informative as any other, or as misinformative (cf. Spiegelhalter et al., 1994, p. 364). It is true that in suitable circumstances accumulating data causes the likelihood to dominate the prior distribution fairly quickly, so that a uniform prior (like Lindley's) can be used as a reasonable approximation to the 'true' one. This is the phenomenon of 'stable estimation' described by Edwards et al. (1963), and further refined by among others Lindley. That is not the point, though. The discussion is not about computationally useful approximations, but about a definite recommendation for choosing priors in general, and making one recommendation in particular: the prior distribution should be 'informationless'. But as we see, the recommendation is impossible to grant, in principle.

Behind it often lies a confusion between *reporting an experiment* and *evaluating a hypothesis* on its evidence. Box and Tiao do so in a single sentence:

> Even when a scientist holds strong prior beliefs about the value of a parameter θ, nevertheless, in reporting his results it would usually be appropriate and most convincing to his colleagues if he analyzed the data against a *reference* prior which is dominated by the likelihood. (1973, p. 22)

In reporting an outcome, the data should certainly speak for themselves. But as we saw, they do not speak for themselves with a reference prior or with any other. Each choice of prior will interpret the data in a different way. There is nothing informationless about that. *There are no informationless priors.* That being the case, we should presumably try to ensure that the priors we do in fact adopt are reasonable ones. Are reference priors reasonable? According to Box and Tiao, the scientist using a reference prior 'could then say that, irrespective of what he or anyone else believed to begin with, the prior distribution represented what someone who *a priori* knew very little about θ should believe in the light of the data' (ibid.). It is difficult to think of worse advice, and advice which, moreover, negates the unique *strength* of Bayesian inference *that it makes explicit allowance for the incorporation of prior evidence*. For the scientist does not usually know very little about θ (we can regard θ as representing any hypothesis here): on the contrary, there is usually a good deal of prior information around, and to deliberately suppress it in a theory which gives it an explicit role is perverse; it is to ignore what that theory is trying to tell you about the correct analysis of data. And what it is trying to tell you is that the information conveyed in the prior is essential information. To replace your evaluation of the prior evidence with a prior that tries to say that there is none is effectively to *distort* what the data should be interpreted as saying. Here is a somewhat unfair example. If the experimental data are that the DNA of the suspect matches that found at the crime, then it looks like strong evidence that the suspect is guilty. But if I know that the suspect was with me throughout any reasonable estimate of the time at which the crime was committed, then (to me) it ceases to be strong evidence, or indeed evidence at all. The example is somewhat unfair because one might respond, and I am sure Box and Tiao would, that of course one should use *all* available data, but one should also let them speak for themselves with a reference prior. But now we come back to the fact that they do not speak for themselves, with any prior. We can also get a less unfair example by going back to the sorts of reason why the mean θ_0 in Lindley's example might merit a reasonable positive probability: it might be a 'true' value predicted by some successful physical theory, say. In that case to adopt a reference prior across the entire range of θ would give a posterior probability of 0 for θ_0. Since we can nearly always represent a theory by a particular value of a continuous parameter, we should have to conclude that no data could ever confirm any theory!

6.2 Simplicity

We have, not entirely accidentally, assembled all the ingredients required for a discussion of that other siren, *simplicity*. We are still in the frame of suggestions for some objective way of choosing, or at any rate constraining, prior probabilities. Simplicity is one frequently suggested, though as a criterion more for *model selection* than for hypotheses within models. But it has its problems. Whether one is a Bayesian or not, the main one is its lack of univocality: there are different ways in which we judge things simple, and these are not all equivalent. But a segment of recent opinion has nevertheless tended to favour the view originally due to the great twentieth-century Bayesian Harold Jeffreys that it is an Occam's razor sense of simplicity that is important to science: to be precise, simplicity considered as *fewness of independent adjustable parameters*. Returning to Jeffreys's modified Galileo law above, which no one of course would accept in preference to Galileo's own, we see that it has the form of Galileo's law with n adjustable parameters evaluated at the data points $t_1, \ldots, t_n$. It is (unnecessarily) more complex than Galileo's simpler law, and we prefer simple laws. Granted this (though I shall question it later), why not elevate this insight into a rule for constraining prior probabilities, that simpler hypotheses in the paucity-of-parameters sense receive greater prior probability? Why not follow Jeffreys and adopt his Simplicity Postulate (1961, pp. 46–50)?

Forster and Sober answer the question by claiming that to do so is *impossible*. Their argument is that since a linear relation is also a polynomial of every higher degree with the coefficients of all terms of degree greater than one set equal to zero, the linear hypothesis cannot have a larger probability than, say, a parabolic since the first entails the second and probability must respect entailment (1994; the charge is repeated in Sober, 1997, where it is called a 'fundamental problem facing Bayesianism'). Their argument can be rebutted simply by noting that the interest is usually in testing against each other not compatible but *incompatible* hypotheses, for example whether the data are better explained by the existing hypothesis or by adding a new parameter in the form of a *non-zero* coefficient to a higher-degree term. Thus, suppose *LIN* is the set of all linear models and *QUAD* the set of all quadratic ones. In testing whether the true model is a linear one M_L or a quadratic one M_Q the tester is *not* testing *LIN* against *QUAD*; since they have common elements it would be like testing M_L against M_L. The test is between *LIN* and *QUAD** where $QUAD* = QUAD \cap LIN^C$ and LIN^C is the set-theoretical complement of *LIN*. While $P(LIN)$ is necessarily no greater than $P(QUAD)$ by the probability calculus, $P(LIN)$ may consistently be greater than $P(QUAD*)$. Whether it should be as a matter of principle is another matter, which I shall come to shortly.

Jeffreys himself regarded tests between such disjoint families as *LIN* and *QUAD** as the classic arena for the Simplicity Postulate, pointing out that the penalty of being able to fit the data exactly by means of a plentiful enough supply of free parameters is overfitting: 'If we admitted the full n [parameters] ... we should change our law with every observation. Thus the principle that laws have some validity beyond the original data would be abandoned' (1961, p. 245). Despite this reasoned defence (indeed, reasoned from the very same first principles they themselves adopt), Forster and Sober, amazingly, call the restriction of the simplicity ordering to disjoint polynomial families an 'ad hoc maneuver' which '*merely changes the subject*' (1994, p. 23; their italics). I think enough has been said to show that the charge is unfounded.

The hypothesis that some relationship is a particular degree of polynomial is to say that it is some (unspecified) member of the corresponding family of curves, and computing its posterior probability means computing the posterior probability, and hence the likelihood, of that family. On this point Forster and Sober make yet another strange statement: 'it remains to be seen how [Bayesians] ... are able to make sense of the idea that families of curves (as opposed to single curves) possess well-defined likelihoods' (1994, p. 23). With respect, it does not remain to be seen at all. The formal technique for 'making sense of the idea' has been around for over two centuries, and it is nothing more than Bayes's Theorem plus integration. Chapter 5 of Jeffreys's book gives an example which happens, serendipitously, to be formally very similar to the one in Lindley's Paradox: the null hypothesis H_0 is that a suggested additional parameter (ϕ) is not needed (i.e. $\phi_0 = 0$), and the alternative, H_1, is that it is and takes some non-zero value (we might for example be testing a linear hypothesis against the alternative that the relationship is some proper quadratic). Assuming as Jeffreys does that the error is normally distributed, we obtain the same equations as in the Lindley example, with the likelihood of H_1 equal to $\int \sqrt{(2\pi\sigma^2)}\exp[-(1/2\sigma^2)(a - \phi)^2]f(\phi)d\phi$ where $f(\phi)$ is a prior density and a is the maximum-likelihood estimate of ϕ (in Lindley's example a is the sample mean); the integration is over the full range of values of ϕ (1961, 5.0). This likelihood is certainly well defined. Admittedly the prior densities Jeffreys uses are usually improper, but (a) that is hardly a necessary choice, and (b) the likelihoods are still well defined even with improper priors, since likelihoods are determined only up to proportionality. The priors may of course be subjective, but that still does not make the likelihoods *ill defined*; it merely gives them a subjective tinge.

There are consistency problems, however, arising from an unrestricted use of the Simplicity Postulate in the Bayesian Theory and a consistent employment is very difficult (Jeffreys himself does not use it consistently; see Howson, 1988). Forster and Sober attempt by contrast a justification for a preference for simpler theories within classical estimation theory: simpler theories (still in Jeffreys's sense) have, they claim, greater *expected predictive accuracy*. In support of this claim they cite Akaike's well-known result that $l(m) - m$ is an unbiased estimate[10] of the expected fit (likelihood) of a model $H(m)$, where m is the number of adjustable parameters and $l(m)$ is the log-likelihood of the maximum of the likelihood function for the family determined by $H(m)$. What has captured some people's imaginations (including those of Forster and Sober) is that the currently best-fitting hypothesis automatically carries a penalty proportional to the number of its free parameters. If maximum likelihood by itself were taken as the guide, then the highest-dimensional hypothesis would always be chosen, with the almost inevitable consequence that it would overfit future data; Akaike's criterion appears to correct for overfitting in a principled way.

But those principles are highly questionable, couched as they are simply in terms of expectations. What is the value of knowing an expected value? And that is merely at one remove: we do not even know it but can only estimate it using an estimator whose justification is that its own expected value is the expected value in question. Bearing in mind also that it is a simple matter to construct *uncountably* many unbiased estimators of a quantity τ which will differ from any specified unbiased estimator on the given data by as much as you like, Akaike's criterion arguably tells us nothing whatever about the relation between the data we have and the true model. Criteria based on expectations allegedly have their home in repeated-sampling theory, but even there the fact that we are doomed to be witness only to the single case remains a barrier to any inference from the data that is not some form of hand-waving (a large sample is still a single member in the set of repeated large samples). Classical estimation criteria are poor surrogates for the real quantity of interest, i.e. the probability that the true value lies within any given limits. For that one needs a theory of posterior probability, which only the Bayesian Theory provides. Moreover, there is a Bayesian analysis, developed by Schwarz, of the relation between the dimensionality of the model and posterior probability that issues in a choice criterion similar to Akaike's in subtracting a penalty proportional to the number of parameters (Schwarz, 1978; for a comparison with the Akaike criterion see Stone, 1979, and Smith and Spiegelhalter, 1980).

Forster and Sober believe that such criteria — specifically, for them, Akaike's — are relevant in a more general context of theory choice in science (they consider the competing models of Ptolemy and Copernicus, 1994,

---

[10] A statistic T is said to be an unbiased estimator of a parameter τ if T's expected value exists and is equal to τ.

pp. 14–15). I doubt this, and for reasons which have nothing to do with how well founded the Akaike criterion is. First, results, even Bayesian ones, developed within the domain of standard statistical modelling, with its characteristic assumptions of independent samples and determinate probability distributions, are not obviously applicable outside it (cf. Kieseppä, 1997). Second, I doubt that simplicity is actually seen as a criterion of dominating importance anyway, at any rate in developed science. There the principal criterion guiding the choice of explanatory theory seems to be degree of connectedness to suitably selected background information (to take a celebrated example, Special Relativity was virtually derived from two postulates, the Relativity Principle and the constancy of *c*), and a corresponding dislike of *ad hoc* assumptions. Developed science is not like blind curve-fitting: the avoidance of adjustable parameters wherever possible is more likely to arise from a view of them as *ad hoc* than from a desire to avoid overfitting (though of course these considerations are not completely independent). It is not the *number* of adjustable parameters that disqualified the bogus free-fall law above, but the fact that there is no good reason in accepted theory for introducing any of them.[11]

We are back on ground familiar from the discussion of reference priors, and the way they caused prior information to be ignored or — what comes to the same thing — misrepresented. Here too: I may well believe on the basis of what I regard as a well-founded belief that in a particular field the best theory will necessarily have a fairly large number of undetermined parameters; in economic forecasting, for example, no simple hypothesis would be regarded as very plausible. The usual reply to this is that the objection fails to observe the principle that one should condition on *all* the available information, which one does using — of course — a simplicity-based prior. Compare the discussion of reference priors again, where a similar rejoinder was made, with 'reference prior' in place of 'simplicity prior'. The answer here is slightly different, however: it is that most of the 'information' cannot be conditioned on in any obvious way. It may consist of a certain weight being put on analogy, or on general heuristic or metaphysical considerations (Einstein put a low prior weight on Copenhagen quantum mechanics because of the stochastic element it incorporated as a core feature). Of course, one might grant all this but now say that one should choose the simplest prior consistent with such constraints. But that just gives the game away: it is prior *plausibility* that is the ultimate criterion. In a discussion of Goodman's celebrated hypothesis 'All emeralds

---

[11] Edgeworth, responding to Pearson's attempt to fit his Type 1 etc. curves to the data, pointedly asked 'what weight should be attached to this correspondence by one who does not perceive any theoretical reason for those formulas?' (1895).

are grue'[12] as an alternative explanation of a sample of green emeralds, Miller identifies what it is about it that is so objectionable: not that it is not simple; lack of simplicity is not what is wrong with it. What is wrong with it is that 'we all know that "All emeralds are grue" *is false*' (1994, p. 37; my italics).

But now we are right back where we started: people find different things plausible. Who is to judge (this is just the misgiving that Richard Swinburne reports in his introductory essay as regards the 'plausibility' view)? The answer, I believe, is 'nobody'. To see why this is the right answer we need to add in some general *philosophical* considerations. An argument Hume enunciated two and a half centuries ago and which is now broadly accepted is that there is a virtual infinity of different explanatory theories consistent with the data, and any attempt to narrow down the field, even to the mere extent of saying that some are more probable than others, will necessarily beg the question. This suggests that the indeterminacy of the priors in the Second Bayesian Theory is neither an accidental nor an objectionable feature: it is an implicit recognition of a *fact*.

This view would be corroborated if it could be shown that the probability calculus is as it stands a *complete* theory of valid probabilistic reasoning. Ramsey viewed the calculus as rules of consistency for epistemic probability assignments, though he never proved that it was a complete set, nor is it clear that he could have done given what he understood by consistency (I shall come to that shortly). Nowadays saying that certain rules are rules of consistency is taken to mean that they are rules of logic, and that gives us a lead. We know for deductive logic what completeness means, and we even have a provably complete system in first-order logic. This prompts two questions: (a) whether there is an authentically logical view of probability with the rules of probability regarded as consistency constraints, and (b) whether a completeness theorem can be proved for them. The answer is 'yes' to both questions, as I shall now show.

II

1. The Second Bayesian Theory as logic

The Second Bayesian Theory, invented independently in the 1920s and 1930s by Ramsey in England and de Finetti in Italy, is not now usually thought to

---

[12] An emerald is defined by Goodman (1946) to be grue if it is observed up to and including now and is green, or is observed after now and is blue. Goodman's construction is clearly a qualitative analogue of the bogus free-fall law.

be a part of logic, though arguably both its founders regarded it as such. Ramsey certainly did and de Finetti used language that seems to indicate a similar view. But the idea that epistemic probability is an extension of deductive logic was already a very old one, going right back to the beginnings of the mathematical theory in the seventeenth century. Leibniz in the *Nouveaux Essais* and elsewhere said so explicitly, and the idea runs like a thread, at times more visible, at times less, through the subsequent development of the epistemic view of probability. Strangely enough, Ramsey did more than anyone else to deflect it from a logical path by choosing to embed his discussion not within the theory of logic as it was then being (very successfully) developed on the Continent, starting with Frege and continuing through Hilbert, Herbrand, Löwenheim, Skolem and Gödel, but within a theory of *utility*. His achievement was remarkable, inaugurating the current paradigm for decision theory, but it put epistemic probability in a setting where its logical character is effectively obscured. Ramsey's talk of consistent preferences is more loose folk usage than anything to do with logical consistency. Indeed, most people have preferred to see Ramsey's, Savage's etc. axioms as very general rationality constraints, and rationality is not logic.

Nevertheless it can be demonstrated, and I shall do so, that the probability axioms really are consistency constraints in an authentically logical sense, and that moreover they are complete, in a sense like that in which first-order logic is complete. To this end, recall that from time immemorial it has been the custom to use odds as the measure of uncertainty,[13] and people have been careful to contrast what they estimate in the light of their background beliefs as the 'true' or 'just' odds with those that in a bet would unfairly advantage one side. Whether any absolute meaning can be given this notion is not something I want to go into here (I think not, for similar sorts of reasons to those I used earlier in talking about objective prior probabilities); what I am interested in now is only the structural properties such odds obey. The odds on a proposition A are of course the ratio k of what is paid if A is false, say R, for what is gained if A is true, say Q; i.e. $k = R/Q$. Since the 'true' or 'just' odds are just the agent's evaluation of the ratio of the corresponding chances for and against A's truth, the bet will be a fair one, i.e. not advantage unfairly one side or the other, if the odds are the agent's 'true' odds.

The odds scale is not a good one on which to measure uncertainty, as it is infinite in the positive direction and unbalanced in that even money odds are

---

[13] E.g. PROTEUS: But now he parted hence, to embark for Milan.

      SPEED: Twenty to one, then, he is shipp'd already. (William Shakespeare, *Two Gentlemen of Verona*)

located very near one end and infinitely far from the other. So people instead used *normalized odds*, i.e. the quantity $p = R/(R+Q) = k/(k+1)$, the so-called *fair-betting quotient*, as the measure of uncertainty. From this we obtain the reverse identity $k = p/(1-p)$, where $0 \leq p \leq 1$, and a familiar formal characterization of a fair bet as one where the odds are the agent's true odds: if the agent's fair-betting quotient is p and the odds in the bet are R/Q, then the bet is fair, in the advantage-equilibrating sense above, just in case $R/Q = p/(1-p)$, i.e. just in case $Qp - R(1-p) = 0$; i.e. *just in case the expected value is zero*.

Suppose the bet with stake $S = R + Q$ is fair. Its payoff table can be represented in the convenient form that de Finetti made familiar:

$$
\begin{array}{ll}
 & A \\
T & S(1-p) \\
F & -pS
\end{array}
$$

Where I_A is the indicator function of A, the bet can therefore be expressed as a random quantity $S(I_A - p)$; this will be useful later. Besides ordinary odds there are also *conditional odds*, in bets on a proposition A which require the truth of some proposition B for the bet to go ahead. The betting quotient in such a conditional bet is called a conditional betting quotient. A conditional bet on A given B with stake S and conditional betting quotient p clearly has the form $I_B S(I_A - p)$.

Let us now think of what it means for these assignments to be *consistent*. There is already at hand a well-known notion of consistency for assignments of numbers to compounds involving number variables, and this is *equation consistency*, or *solvability*. A set of equations is consistent (solvable) if there is at least one assignment of values to its variables which does not overdetermine any of them. Although this aspect of it is not stressed in the usual logic texts, deductive consistency itself is really nothing but solvability in this sense. This may seem surprising, because consistency is usually seen as a property of sets of sentences. However, it is not difficult to see that we can equivalently regard consistency as a property of truth-value assignments. First, note that according to the classical Tarskian truth definition for a first or higher-order language conjunctions, disjunctions and negations are homomorphically mapped onto a Boolean algebra of two truth values, {T, F}, or {1, 0}, or however they are to be signified. Now consider any attribution of truth values to some set Σ of sentences of L, i.e. any function from Σ to truth values. Call any such assignment CONSISTENT if it is capable of being extended to a single-valued function from the entire set of sentences of L to truth values which satisfies those homomorphism constraints. The theory of 'signed' semantic tableaux or trees is a syntax adapted to such a way of looking at, and testing for, CONSISTENT assignments. ('Signing' a tableau just means appending Ts and Fs to the constituent sentences. The classic treatment is Smullyan, 1968, pp. 15–30; a simplified account is in Howson, 1997.) Here is a very simple example:

$$A \; T$$
$$A \to B \; T$$
$$B \; F$$

The tree rule for $[A \to B \; T]$ is the binary branching

$$/ \quad \backslash$$
$$F A \quad B \; T$$

Appending the branches beneath the initial signed sentences results in a *closed tree*, i.e. one on each of whose branches occurs a sentence to which is attached both a T and an F. A soundness and completeness theorem for trees (Howson, 1997, pp. 107–11) tells us that any such tree closes if and only if the initial assignment of values to the three sentences A, $A \to B$, and B is inCONSISTENT, i.e. unsolvable over L subject to the constraints of the general truth-definition.

To see that CONSISTENCY and consistency are essentially the same concept, note that an assignment of truth values to a set Σ of sentences is CONSISTENT just in case the set obtained from Σ by negating each sentence in Σ assigned F is consistent in the standard (semantic) sense. In algebraic treatments of logic the identity becomes more apparent: to show that a set of propositional formulae is consistent is to show that as a system of simultaneous Boolean polynomial equations (equated to 1, the maximal element of the Boolean algebra) it has a solution over the propositional variables. Thus in deductive logic (semantic) consistency can be equivalently defined in the equational sense of a truth-value assignment being solvable, i.e. extendable to a valuation over all sentences of L satisfying the general rules governing truth valuations, and we can call such an extension a *model* of the initial assignment.

The variables evaluated in terms of betting quotients are traditionally propositions and not sentences, but this is no great formal dissimilarity since propositions can be regarded simply as equivalence-classes of sentences, with an obvious rule for distributing the probability to the constituent sentences. By analogy with deductive CONSISTENCY we can say that an assignment of fair-betting quotients is consistent just in case it can be solved in a analogous sense, by being extendable to a single-valued assignment to all the propositions in the domain of discourse, or *language L* (cf. Paris, 1994, p. 6), subject to suitable constraints analogous to the Tarskian ones in the deductive case. What should those constraints be? We are talking about *fair*-betting quotients (in the agent's estimation at any rate), and just as the Tarskian conditions characterize truth-in-general, so the constraints here should be those formally characterizing the *purely general* content of the notion of fairness: call them (F). The odds expressing the strength of your beliefs about A are fair (relative to that estimation) if they give no calculable advantage to either side. What this implies in the way of completely general constraints seems to amount to:

(a) If p is the fair-betting quotient on A, and A is a logical truth, then $p = 1$; if A is a logical falsehood, $p = 0$.
(b) Fair bets are invariant under change of sign of stake.

The reason for (a) is not difficult to see. If A is a logical truth and p is less than 1, then in the bet $S(I_A - p)$ with betting quotient p, I_A is identically 1 and so the bet reduces to the positive quantity $S(1 - p)$ received come what may. Hence the bet is not fair since one side has a manifest advantage. Similar reasoning shows that if A is a logical falsehood, then p must be 0. As to (b), changing the sign of the stake reverses the bet, and according to (F) the fair-betting quotient confers no advantage to either side, from which (b) follows immediately.

Do (a) and (b) exhaust the content of (F)? Not quite: there is something else besides, a natural closure condition which says that your views of what betting quotients are fair must, as betting quotients, respect the structural relations between bets. In particular:

(Closure)   *If a finite or denumerable sum of fair bets determines a bet on a proposition B with betting quotient q, then q is the fair betting quotient on B.*

We spoke of the constraints determined by (F) as analogous to those determined by the general truth conditions of deductive logic. And indeed there is a closure condition there too, implicit in the fact that truth satisfies the usual truth-table definitions of the connectives (and of course the satisfaction conditions for the quantifiers). But suppose you entertain a correspondence conception of truth. The standard basis condition of the Tarskian truth definition will not worry you, saying as it does that an atomic sentence $B(a)$ is true just in case the individual denoted by a has the predicate denoted by B. That seems a clear case of correspondence. But it is not so obvious what sort of fact corresponds to the truth of a sentence $-A$. Are there really negative facts? (This question did indeed preoccupy some philosophers.) Nevertheless, even for correspondence theorists, of whom Tarski himself was explicitly one, truth is now accepted as obeying all the inductive clauses of the usual (classical) truth definition. So accepted is it that it is seldom if ever now questioned. Yet it is a very powerful condition: practically all of what we understand by deductive logic follows from it.

So too here. To see how powerful it is, note that compound bets obey the following arithmetical conditions:

(i) $-S(I_A - p) = S(I_{-A} - (1-p))$.

(ii) If $A\&B = \bot$, then $S(I_A - p) + S(I_B - q) = S(I_{A\lor B} - (p + q))$.

(iii) If $\{A_I\}$ is a denumerable family of propositions in $B(F)$ (see next paragraph) and $A_i\&A_j = \bot$, and p_i are corresponding betting quotients and Σp_i exists, then $\Sigma S(I_{Ai} - p_i) = S(I_{\lor Ai} - \Sigma p_i)$.

(iv) If $p, q > 0$, then there are non-zero numbers S, T, W such that $S(I_{A\&B} - p) + (-T)(I_B - q) = I_B W(I_A - p/q)$ (T/S must be equal to p/q). The right-hand side is clearly a conditional bet on A given B with stake W and betting quotient p/q.

Closure tells us that if the betting quotients on the left-hand side are fair, then so are those on the right. The way the betting quotients on the left combine to give those on the right is, of course, just the way the probability calculus tells us that probabilities combine over compound propositions and for conditional probabilities. Now for the central definition. Let Q be an assignment of personal fair-betting quotients to a subset X of $B(F)$, where $B(F)$ is the Borel field generated by the atomic quantifier-free sentences of some first-order language L (Paris, 1994, pp. 164–71). By analogy with the deductive case, we shall say that Q is *consistent* if it can be extended to a single-valued function on all the propositions of L satisfying suitable constraints, in this case the general conditions of fairness including closure. It is now a short step to the following:

Theorem: An assignment Q of fair-betting quotients (including conditional fair-betting quotients) is consistent if and only if Q satisfies the constraints of the countably-additive probability calculus. (The proof is straightforward; see Howson, 2000, pp. 130–32.)

I pointed out that there is a soundness and completeness theorem for first-order logic which establishes an extensional equivalence between a semantic notion of consistency, as a solvable truth-value assignment, and a syntactic notion, as the openness of a tree from the initial assignment. In the theorem above we have an analogous soundness and completeness theorem for a quantitative logic of uncertainty, establishing an extensional equivalence between a semantic notion of consistency, i.e. having a model, and a syntactic one, deductive consistency with the probability axioms when the probability functor P signifies the fair-betting quotients in Q. The Bayesian theory is an authentic *logic*.

CONCLUSION

The result above is a vindication of Hume's argument (I have argued this at greater length in Howson, 2000). The probability axioms are the *complete* logic of probable inference, whose only absolute assignments are to trivial propositions. The situation is the same as with deductive logic: to get information out you need to put it in, and as in deductive logic what you put in as a premise will be at least as fallible and conjectural as what you get out as a validly derived conclusion. That is on the negative side; it just amounts to what most philosophers have reluctantly come to accept as fact, that the problem of induction is insoluble. On the positive side we now have a theory of valid probable inference, as much a logic as deductive (first-order) logic, and with a corresponding soundness and completeness theorem. That is no mean achievement. In addition, and vindicating the beliefs of the seventeenth- and eighteenth-century pioneers, statistical inference and non-statistical inductive inference become jointly subsumed under a uniform set of principles. Fisher himself also believed that they should be, but took as his model of scientific inference, or of a substantial part of it, a hypothetico-deductive one, and though hypotheses about frequency distributions do not deductively entail any observation statement, he determined, Procrustes-like, that they should be made to by cutting off not their feet but their tails. The result does not behave properly, as might be expected from the anatomical analogy, and as Lindley's result above convincingly proves.

Note. I would like to thank Donald Gillies, John Howard, Larry Phillips, Peter Urbach, and particularly Dennis Lindley, who all read the manuscript and gave much helpful advice.

References

Barnard, G. A. (1990), 'Must Clinical Trials be Large? The Interpretation of P-Values and the Combination of Test Results', *Statistics in Medicine* **9**: 601–14.

Bayes, T. (1763), 'An Essay Towards Solving a Problem in the Doctrine of Chances', *Philosophical Transactions of the Royal Society* **53**: 370–418. Reprinted in this volume as Appendix, pp. 122–49.

Berger, J. O. and Delampady, M. (1987), 'Testing Precise Hypotheses', *Statistical Science* **2**: 317–52.

Berger, J. O. and Sellke, T. (1987), 'Testing a Point Null Hypothesis: the Irreconcileability of P Values and Evidence', *Journal of the Americal Statistical Association* **82**: 112–22.

Birnbaum, A. (1962), 'On the Foundations of Statistical Inference', *Journal of the American Statistical Association* **57**: 269–306.

Box, G. E. P. and Tiao, G. C. (1973), *Bayesian Inference in Statistical Analysis*, Reading, MA: Addison-Wesley.

Casscells, W., Shoenberger, A. and Grayboys, T. (1978), 'Interpretation by Physicians of Clinical Laboratory Results', *New England Journal of Medicine* **299**: 999–1000.

De Finetti, B. (1937), 'Foresight, Its Logical Laws, Its Subjective Sources', translated and reprinted in *Studies in Subjective Probability*, eds H. Kyburg and H. Smokler, New York: Wiley, 1964, pp. 93–159.

Earman, J. (1992), *Bayes or Bust: A Critical Examination of Bayesian Confirmation Theory*, Cambridge, MA: MIT Press.

Edgeworth, F. Y. (1895), 'On Some Recent Contributions to the Theory of Statistics', *Journal of the Royal Statistical Society* **58**: 505–15.

Edwards, A. W. F. (1970), *Likelihood*, Cambridge: Cambridge University Press.

Edwards, W., Lindman, H. and Savage, L. J. (1963), 'Bayesian Statistical Inference for Psychological Research', *Psychological Review* **70**: 193–242.

Fisher, R. A. (1926), *Statistical Methods for Research Workers*, Edinburgh: Oliver and Boyd (references to 1932 edition).

Fisher, R. A. (1935), *The Design of Experiments*, Edinburgh: Oliver and Boyd.

Fisher, R. A. (1956), *Statistical Methods and Statistical Inference*, Edinburgh: Oliver and Boyd.

Forster, M. and Sober, E. (1994), 'How To Tell When Simpler, More Unified or Less Ad Hoc Theories Will Provide More Accurate Predictions', *British Journal for the Philosophy of Science* **45**: 1–35.

Freeman, P. (1993), 'The Role of P-Values in Analysing Trial Results', *Statistics in Medicine* **12**: 1443–58.

Good, I. J. (1983), *Good Thinking*, Minneapolis: University of Minnesota Press.

Goodman, N. (1946), *Fact, Fiction, and Forecast*, New York: Bobbs-Merrill.

Howson, C. (1988), 'On the Consistency of Jeffreys's Simplicity Postulate, and its Role in Bayesian Inference', *The Philosophical Quarterly* **38**: 68–83.

Howson, C. (1997), *Logic With Trees*, London: Routledge.

Howson, C. (2000), *Hume's Problem: Induction and the Justification of Belief*, Oxford: Oxford University Press.

Howson, C. (2001), 'A New Kind of Logic', in David Corfield and Jon Williamson (eds), *The Foundations of Bayesianism*, Dordrecht, The Netherlands: D. Reidel.

Howson, C. and Urbach, P. M. (1993), *Scientific Reasoning: the Bayesian Approach*, 2nd edn, Chicago: Open Court.

Jeffreys, H. (1961), *Theory of Probability*, 3rd edn, Oxford: Oxford University Press.

Kadane, J. B., Schervish, M. J. and Seidenfeld, T. (1996), 'When Several Bayesians Agree that there will be no Reasoning to a Foregone Conclusion', *Proceedings of the Biennial Meeting of the Philosophy of Science Association*, Part 1, 281–90.

Kendall, M. and Stuart, A. (1973), *The Advanced Theory of Statistics*, London: Charles Griffin & Co., vol. 2.

Kieseppä, I. A. (1997), 'Akaike Information Criterion, Curve-fitting, and the Philosophical Problem of Simplicity', *British Journal for the Philosophy of Science* **48**: 21–49.

Lee, P. M. (1997), *Bayesian Statistics*, 2nd edn, London: Arnold.

Lindley, D. V. (1957), 'A Statistical Paradox', *Biometrika* **44**: 187–92.

Miller, D. (1994), *Critical Rationalism: A Restatement and Defence*, Chicago: Open Court.

Paris, J. (1994), *The Uncertain Reasoner's Companion*, Cambridge: Cambridge University Press.

Phillips, L. D. (1973), *Bayesian Statistics for Social Scientists*, Cambridge: Thomas Y. Crowell Press.

Ramsey, F. P. (1926), 'Truth and Probability', *The Foundations of Mathematics and Other Logical Essays*, London: Routledge and Kegan Paul, 1931, pp. 156–99.

Schwarz, G. (1978), 'Estimating the Dimension of a Model', *Annals of Statistics* **6**: 461–4.

Smith, A. F. M. and Spiegelhalter, D. J. (1980), 'Bayes Factors and Choice Criteria for Linear Models', *Journal of the Royal Statistical Society* B **42**: 213–20.

Smullyan, R. M. (1968), *First Order Logic*, New York: Dover.

Sober, E. (1997), 'What is the Problem of Simplicity?', forthcoming.

Spiegelhalter, D. J., Freedman, L. S. and Parmar, M. K. B. (1994), 'Bayesian Approaches to Randomised Trials', *Journal of the Royal Statistical Society* A **157**: 358–416.

Stone, M. (1979), 'Comments on Model Selection Criteria of Akaike and Schwarz', *Journal of the Royal Statistical Society* B **41**: 276–8.

4

Bayes's Theorem and Weighing Evidence by Juries

A. P. DAWID

1. Statistics and the law

1.1 Evidence

AT FIRST SIGHT, there may appear to be little connection between Statistics and Law. On closer inspection it can be seen that the problems they tackle are in many ways identical — although they go about them in very different ways. In a broad sense, each subject can be regarded as concerned with the *interpretation of evidence*. I owe my own introduction to the common ground between the two activities to my colleague William Twining, Professor of Jurisprudence at University College London, who has long been interested in probability in the law. In our discussions we quickly came to realize that, for both of us, the principal objective in teaching our students was the same: to train them to be able to interpret a mixed mass of evidence. That contact led to my contributing some lectures on uses and abuses of probability and statistics in the law to the University of London Intercollegiate LLM course on Evidence and Proof (and an Appendix on 'Probability and Proof' to Anderson and Twining (1991)), as well as drawing me into related research (Dawid, 1987, 1994; Dawid and Mortera, 1996, 1998). To my initial surprise, I found here a rich and stimulating source of problems, simultaneously practical and philosophical, to challenge my logical and analytical problem-solving skills. For general background on some of the issues involved, see Eggleston (1983); Robertson and Vignaux (1995); Aitken (1995); Evett and Weir (1998); Gastwirth (2000).

The current state of legal analysis of evidence seems to me similar to that of science before Galileo, in thrall to the authority of Aristotle and loth to concede the need to break away from old habits of thought. Galileo had the revolutionary idea that scientists should actually look at how the world behaves. It may be equally revolutionary to suggest that lawyers might look at how others have approached the problem of interpretation of evidence, and that they might

Proceedings of the British Academy, **113**, 71–90. © The British Academy, 2002.

even have something to learn from them. It is my strong belief (though I do not expect it to be shared by many lawyers) that statisticians, with their training in logical analysis of the impact of evidence on uncertainty, have much to contribute towards identifying and clarifying many delicate issues in the interpretation of legal evidence. And I believe that by far the greatest clarity and progress will come from the exploitation of the Bayesian philosophy and methodology of statistics. Although other statistical approaches are possible, I consider that these are at best unnecessary and at worst dangerously misleading.

My message is not a new one. Some of the earliest work on probability (by such great minds as James and Nicholas Bernoulli, Condorcet, Laplace, Poisson, Cournot) was concerned with, or motivated by, problems of quantification and combination of legal evidence and judgement. However, both statisticians and lawyers seem to have lost this thread. I hope I can persuade both communities that it is well worth picking up and following.

2. Testing between two hypotheses — in court

In a criminal case, each charge on the indictment sheet can be regarded as a hypothesis under test. For simplicity we restrict attention to the case of a single defendant and a single charge against that defendant. The proposition that the defendant is *guilty* of the charge (a proposition which typically combines issues both of fact and of law) is the *Prosecution Hypothesis*; we shall henceforth denote it by G. It is specific and clearly defined — what statisticians would call a 'simple hypothesis'.

The task of the Defence is to cast doubt on G, or, what is equivalent, to argue for the reasonableness of the contrary proposition $\overline{G}$, which asserts the falsity of G. We call $\overline{G}$ the *Defence Hypothesis*. It can be very non-specific — a 'composite hypothesis', in statistical parlance. The Defence is not obliged to identify or argue in favour of any more specific alternative hypothesis to G, although this will often be appropriate. In that case, $\overline{G}$ might also reduce to a simple hypothesis.

Let us denote the conjunction of one or more items of evidence (perhaps the totality of the evidence) in the case by E. Our task is to use the evidence E to cast light on the two hypotheses, G and $\overline{G}$, before the court.

2.1 Bayesian approach

Since we start out with uncertainty about the hypotheses and the evidence, we should try to quantify that uncertainty appropriately. A vast body of thoughtful logical and philosophical analysis of uncertainty has concluded that the only appropriate quantification of uncertainty, of any kind, is in terms of the *Calculus of Probability*. This is the basis of the modern Bayesian approach to statistical inference. An important aspect of this philosophy is that complete objectivity is an illusion, and thus that there is no such thing as 'the' probability of any uncertain event — rather, each individual is entitled to his or her own subjective probability. This is not, however, to say that anything goes: in the light of whatever relevant evidence may be available, certain opinions will be more reasonable than others. In simple statistical problems, it can be shown that differing initial subjective distributions will be brought into ever closer and closer agreement when updated through the incorporation of sufficiently extensive observational evidence. In legal applications the conditions for this convergence may not apply, but even so there will be certain probabilistic ingredients and conclusions that can be regarded as reasonable by all reasonable parties.

From the standpoint of any individual juror, the end-point of his (or her) analysis of the evidence E heard should be his *posterior probability* of guilt given the evidence:

$$P(G|E) \tag{1}$$

(where we use the notation $P(A|B)$ to denote the conditional probability of an uncertain event A, in the light of known or hypothesized information B). This is a direct expression of the juror's remaining uncertainty as to the validity of the Prosecution Hypothesis, after taking into account the evidence E. However, it will not typically be appropriate to assess this directly; rather, it should be constructed out of other, more basic and defensible, primitive ingredients. This can be done using the 'odds' form of Bayes's Theorem:

$$\frac{P(G|E)}{P(\overline{G}|E)} = \frac{P(G)}{P(\overline{G})} \times \frac{P(E|G)}{P(E|\overline{G})}, \tag{2}$$

or, in words:

Posterior odds = Prior odds × Likelihood ratio.

The first term on the right-hand side of (2), the prior odds, measures the relative degrees of belief as between the Prosecution and the Defence Hypotheses, *before* the evidence E has been incorporated. There may be a range of reasonable values that jurors could hold for this in the light of previous evidence. The second term, the likelihood ratio, *LR*, involves the probabilities accorded to the new evidence *E* by each of the two competing hypotheses. In some cases (essentially when both *G* and $\overline{G}$ can be framed as simple hypotheses) these ingredients will be moderately 'objective' and agreed upon, and thus so also will be the likelihood ratio. It might then seem appropriate to present the value of *LR*, or its constituent probabilities, as a summary of the impact of the evidence, leaving individual jurors to combine it with their own prior beliefs, using equation (2). Once the posterior odds, say Ω, has been calculated, the desired posterior probability of guilt, $P(G|E)$, is just $\Omega/(1 + \Omega)$. (When one or both of the hypotheses is composite, more complex calculations may be required, incorporating prior assessments even into the calculation of the likelihood ratio.)

Simple though it is, both the logic and the calculation involved in equation (2) will be beyond most judges (and even some jurors). It could then be helpful to present the impact of the evidence by means of a table, incorporating the relevant value of *LR*, and showing how any hypothetical prior probability of guilt, $P(G)$, would be updated by the evidence to the posterior probability $P(G|E)$. If the resulting posterior probability is sufficiently high (and it would be up to the Judge to advise on what this means), then the verdict 'Guilty' would be appropriate. Table 1 shows the impact of a likelihood ratio value of 100 (i.e. the probability of the evidence *E* is 100 times greater under the Prosecution Hypothesis than it is under the Defence Hypothesis). If it is appropriate to regard satisfaction as to guilt 'beyond reasonable doubt' as attained when the posterior probability $P(G|E)$ exceeds 99 per cent, then we see that a juror should be willing to convict whenever his prior probability, before incorporating *E*, exceeded 50 per cent. (Note, however, that when *E* is the only evidence in the case, before *E* is admitted the suspect should be treated no differently from any other member of the population, so that a prior probability of even 0.001 could be regarded as unreasonably high.)

The application of equation (2) can be done either 'wholesale', with *E* representing the totality of the evidence in the case; or 'piecemeal', with sequential incorporation of several items of evidence. In the latter case, the probabilities on the right-hand side should be regarded as evaluated conditional on all previously incorporated evidence. This evaluation might have been done formally, using previous applications of (2) to account for earlier items of evidence, or more informally, if the previous evidence seems too

Table 1. The impact of a likelihood ratio $LR = 100$

Prior prob.	0.001	0.01	0.1	0.3	0.5	0.7	0.9
Post. prob.	0.09	0.50	0.92	0.98	0.99	0.996	0.999

'soft' to justify that. One has to be particularly aware of the possibility that this incorporation of previous evidence might alter the probabilities constituting the likelihood ratio. This will happen if, conditional on the relevant hypothesis, the new item E is not independent of previous evidence. But quite frequently this independence assumption is justified, and then the LR value is unaffected by the previously incorporated evidence.

2.2 Neyman–Pearson and Bayes

The Bayesian approach to statistical inference has been the subject of controversy for at least 150 years, at times almost being consigned to oblivion, and at others (increasingly so in modern times) being strongly in the ascendant. Other schools of statistical inference have grown up which purport to banish subjectivity by using probability in more indirect ways. The most successful of these is the Neyman–Pearson approach, which assesses and selects decision rules by examining their probabilistic performance under various hypotheses. We introduce and examine this in the context of a real court case with which the author was involved.

2.2.1 Regina v. Sally Clark

Sally Clark's first child Christopher died unexpectedly at the age of about three months, when Sally was the only person in the house. The death was initially treated as a case of Sudden Infant Death Syndrome (SIDS, or 'Cot Death'). Sally's second child Harry was born the following year. Harry died in similar circumstances at the age of two months. Soon after, Sally was arrested and tried for murdering both children. There was some medical evidence suggesting smothering, although this could also be explained as arising from attempts at resuscitation. For the moment we disregard the medical evidence, but return to this at § 2.4 below.

At trial, a professor of paediatrics testified that the probability that, in a family like Sally's, two babies would both die of SIDS was around 1 in 73 million. This was based on a professionally executed study which estimated the probability of a single SIDS death in such a family at 1 in 8500 — which figure was then squared by the witness, to account for two deaths. Although this calculation is extremely dubious, being based on unrealistic assumptions of independence, I will accept it here, simply for the sake of argument.

Sally Clark was convicted of murder. Although we cannot know how the jury regarded the statistical evidence, it is reasonable to speculate that it was strongly influenced by the extremely small probability value of 1 in 73 million that both deaths could have arisen from SIDS, regarding this as ruling out the possibility of death by natural causes.

2.2.2 *Neyman–Pearson approach*

Let us see how a Neyman–Pearson statistician might go about making inference from the evidence. For simplicity, suppose that the only possible causes of infant death are SIDS and murder, and discount the possibility that these might both occur in the same family. Consider the decision rule: 'If two babies in a family both die of unexplained causes, decide that their mother murdered them.' Under the hypothesis that a mother does indeed murder two of her babies, the probability of making an error using this rule is 0. Under the contrary hypothesis that she does not, an error will be made if and only if they both die of SIDS, which will occur with probability 1 in 73 million. The rule appears amazingly accurate, under either hypothesis, and thus seems a good one to follow. And, if we apply it to the case of Sally Clark, we must decide that she murdered her babies.

2.2.3 *Criticism*

Even without looking at the above argument through Bayesian spectacles, it does not need much thought to realize that there is something missing. We are considering two hypotheses: death by SIDS, and death by murder. The probability of the former plays a major role in the above Neyman–Pearson argument. But should not the probability of the latter be just as relevant? Office of National Statistics data for 1997 show that, out of a total of 642,093 live births, seven babies were murdered in the first year of life. This yields an estimate of the probability of being murdered, for one child, of 1.1×10^{-5}. If we were to treat this in the same (admittedly dubious) way as the corresponding probability for SIDS, and square it to account for the two deaths, we could argue that the probability that Sally Clark's two babies were both murdered is about 1 in 8.4 billion (we here use the American billion, i.e. one thousand million) — and if this were the argument presented to the jury, how could it but conclude that the possibility of murder is so improbable as to be inconceivable?

Once this is pointed out, it should be clear that the figure quoted of 1 in 73 million for two deaths due to SIDS cannot be significant in itself, but only in relation to the 'counterbalancing probability' of two deaths due to murder.

2.3 Two Bayesian arguments

Only the Bayesian approach can be relied upon to take automatic correct account of all the evidence. However, there can be more than one way of formulating the problem for Bayesian analysis. Here we consider two such formulations, and verify that they yield the same answer.

2.3.1 Restricted hypotheses

Taking into account that the two babies died of something, and that we only consider SIDS and murder as possible causes of death, we can formulate two hypotheses, exactly one of which must be the case:

M: Sally Clark murdered her babies.

S: Sally Clark's babies died of SIDS.

The evidence E before the court is just the fact that the babies died.

By Bayes's Theorem, the posterior odds for comparing the two hypotheses is given by:

$$\frac{P(M|E)}{P(S|E)} = \frac{P(M)}{P(S)} \times \frac{P(E|M)}{P(E|S)}. \qquad (3)$$

The first term on the right-hand side is just the ratio of the overall probabilities of two deaths, under the two hypotheses. Using the hypothetical figures above, this is (1/8.4 billion)/(1/73 million), or about 0.009. As for the second, likelihood ratio, term, E is certain to occur under either hypothesis, so that this is just unity, and the posterior odds is again $0.009 = 9/1000$ — or over $100:1$ against. We conclude that it is over 100 times more probable that the two babies died of SIDS than that they were murdered.

2.3.2 Unrestricted hypotheses

It could be objected that the evidence E has been used twice in the above calculation: once to delimit the hypotheses, and again to form a likelihood ratio. Even though the latter use was completely ineffective, since the likelihood ratio was unity, there might still be a concern because the hypotheses being compared were only formulated in the light of the evidence: in particular, hypotheses which do not lead to the babies' deaths should perhaps not have been excluded, prior to taking the evidence of those deaths into account. To verify that this use of the evidence to refine the hypotheses is not in fact of any importance, we conduct an alternative analysis, in which no such refinement takes place. That is, we now specify the hypotheses as G, that Sally Clark murdered her babies, and $\overline{G}$, that she did not. Then G is identical with M, but $\overline{G}$ is not the same as S, since it does not assume that the babies will die at all — it is just the negation of G.

Once again, Bayes's Theorem, as given by (2), can be applied. We have $P(E|G) = 1$, while $P(E|\overline{G})$ is 1/73 million. The likelihood ratio in favour of guilt is thus 73 million — a seemingly overwhelming weight of evidence in favour of murder. However, we must not forget the first term on the right of (2), the prior odds.

In fact we have $P(G) = P(M) = 1/8.4$ billion; so the prior odds, $P(G)/P(\overline{G})$, is also essentially $1/8.4$ billion. Combining the two factors we obtain once again ($1/8.4$ billion)/($1/73$ million), or 0.009, as the posterior odds on guilt.

It is thus entirely immaterial which of the above two Bayesian arguments is followed. My personal preference in this particular instance is for the former, because it deals symmetrically with the two competing hypotheses about the deaths, thus directly addressing the criticism voiced in § 2.2.3. However, this preference is of no importance so long as the correct Bayesian calculations are performed; and, as we shall see, the latter line of argument is a more natural approach in many problems.

2.4 Additional evidence

In the case of Sally Clark, there was additional medical evidence, including haemorrhages to the children's brains and eyes. Once again, so long as the laws of probability are followed correctly, it does not matter whether this evidence is incorporated before or after the analysis based purely on the fact of death. In this case it is more straightforward to bring it in at the end, as a new likelihood ratio term. Thus suppose (purely for illustration) that the probability of observing the specific medical signs in the case, on the hypothesis that the babies were murdered, is assessed at 1 in 20; while the probability of the same signs being observed on the hypothesis that they died of SIDS can be taken as 1 in 100. The effect of the medical evidence is then to multiply the posterior odds on guilt resulting from the previous analysis by 5. Combined with the hypothetical figures used before, which gave posterior odds of 0.009 based just on the fact of death, this would yield updated posterior odds of 0.043, corresponding to a posterior probability of 4.2 per cent. While the choice of specific numbers here is clearly subject to some vagueness, the Bayesian approach does make it very clear exactly what features of the evidence must be taken into account. In particular, it could be quite misleading to suggest simply that the medical signs observed are 'consistent with' the babies having been smothered. What is required is a quantitative estimate of how many times more likely that evidence would be under the murder hypothesis than it would be under the hypothesis of accidental death followed by attempts at resuscitation.

3. Identification evidence

In many criminal and civil cases, identification of a suspect as the actual perpetrator of an offence is based on an alleged 'match' between trace evidence taken from the scene and a sample taken from the suspect. A fingerprint on the murder weapon is compared with that of the suspect; a footprint found at the scene is compared with shoes belonging to the suspect; the refractive index of fragments of glass found on the suspect's clothing is compared with that of the broken window at the scene of a burglary; fibres left at the scene are compared, by eye, microspectro-fluorimetry, and thin-layer chromatography, with others taken from the suspect's jumper. Increasingly, especially in cases of murder, rape and alleged paternity, the match is based on DNA profiles, and may well be the only evidence presented.

In the following we assume that the fact that a crime has been committed is not in dispute — the only issue is the identity of the culprit.

A forensic scientist will typically summarize the strength of potentially incriminating identification evidence in terms of an associated *match probability*: an assessment of the probability that the observed match would arise if the suspect were in fact innocent. Match-probability values of the order of one in a billion are now routine for DNA profiling. Intuitively, the smaller the match probability, the stronger the evidence against the suspect; and very tiny probabilities appear to be incontrovertibly convincing evidence. However, there are many subtleties and pitfalls in the interpretation of such forensic identification evidence, which are all too often not appreciated.

A typical problem can be formulated as follows. Let M denote the evidence as to the match, and B any other (background) evidence in the case. The full evidence is thus $E = M\&B$. Typically, though by no means universally, it is appropriate to assume that, if the suspect were indeed guilty, then the two samples would be bound to match, and we shall assume this. That is, we take

$$P(M|G) = 1. \qquad (4)$$

The match probability measures how probable the match would be if the suspect were innocent: that is, $P(M|\overline{G})$. Typically this is calculated on the hypothesis that, in this case, the true perpetrator would be entirely unrelated to the suspect. However, this could be misleading if there were any possibility that the suspect and the perpetrator might be related, either closely or distantly. Taking such a possibility into account can greatly affect the value of a DNA profile-match probability.

For the sake of illustration, we shall proceed on the assumption that the match probability is 1 in 10 million:

$$P(M|\overline{G}) = 10^{-7}. \qquad (5)$$

3.1 Speeches in court

3.1.1 The Prosecution argument
Counsel for the Prosecution argues as follows:

> Ladies and gentlemen of the jury, the probability of the observed match between the sample at the scene of the crime and that of the suspect having arisen by innocent means is 1 in 10 million. This is an entirely negligible probability, and we must therefore conclude that, with a probability overwhelmingly close to 1, the suspect is guilty. You have no alternative but to convict.

3.1.2 The Defence argument
Counsel for the Defence argues as follows:

> In the general population, there are about 30 million individuals who might possibly have committed this crime. One of these is the true culprit. Of the remaining 30 million innocent individuals, each has a probability of 1 in 10 million of providing a match to the crimes sample. We should therefore expect there to be three innocent individuals who match. We know that the suspect matches, but he could still be any one of the four matching individuals, of which one is guilty and three are innocent. So the probability that he is guilty is only one quarter. This falls far short of the criminal criterion of 'beyond reasonable doubt'. You must acquit.

3.2 The Bayesian argument

Whatever one thinks of the above two arguments, neither can be generally appropriate, since neither makes any allowance for the incorporation of the other evidence (B) in the case. However, this can readily be done by means of the correct Bayesian argument. We now present this, and then return to reconsider the above Prosecution and Defence arguments.

We start from the position that the other evidence B has already been presented, so that all the probabilities we consider are already conditional on B. We then apply Bayes's Theorem to incorporate, in addition, the match evidence M, so obtaining posterior probabilities given the totality of the evidence, $E = M\&B$. With this understanding, (2) becomes:

$$\frac{P(G|E)}{P(\overline{G}|E)} = \frac{P(G|M\&B)}{P(\overline{G}|M\&B)}$$
$$= \frac{P(G|B)}{P(\overline{G}|B)} \times \frac{P(M|G\&B)}{P(M|\overline{G}\&B)}. \tag{6}$$

We previously assumed (4) that, if the suspect is guilty, then he will certainly provide a match — and this property cannot be affected by learning any further background information. That is, $P(M|G\&B) = 1$.

It will likewise normally be appropriate (particularly for DNA evidence) to disregard the background information when assessing the probability that the suspect would match were he in fact innocent. This will hold when, conditional on innocence, the background evidence and the match identification evidence can be regarded as arising independently. Assuming this in the case at hand, from (5) we have $P(M|\overline{G}\&B) = P(M|\overline{G}) = 10^{-7}$.

The likelihood ratio for the identification evidence, $P(M|G\&B)/P(M|\overline{G}\&B)$, is thus 10^7, unaffected by the other evidence B.

The other term needed for (6) is the prior odds, $P(G|B)/P(\overline{G}|B)$. But, while this is indeed prior to the introduction of the identification evidence, it is posterior to the background evidence B. The initial incorporation of that evidence might have been done formally, by means of an earlier application of Bayes's Theorem, or informally. But in any event it needs to be taken into account in making a realistic assessment of the relevant prior odds. Whatever such assessment is made, the final odds on guilt is obtained by multiplying this prior odds by 10^7.

3.3 Reconsideration of Counsels' arguments

We now look more closely at Defence Counsel's argument.

Let us accept that the appropriate size of the 'catchment area' of potential perpetrators is indeed 30 million. We further suppose that, apart from the DNA evidence, there is no additional background evidence relating to the suspect, other than that he belongs to this catchment population. Then it would be appropriate to take, as the prior probability of guilt, $P(G|B) = 1/30$ million, and the prior odds would be essentially the same — certainly such a judgement would seem to be in the spirit of the 'presumption of innocence'. Applying Bayes's Theorem (6), the posterior odds on guilt are thus (1/30 million) $\times$ (10 million) = 1/3, corresponding to a posterior probability of one quarter. That is to say, the Defence argument agrees with a Bayesian analysis in which the only background evidence used is the size of the population of potential perpetrators. It is implicit that, before the identification evidence is taken into account, anyone in this population is as likely as any other to be the true culprit. In cases where the identification evidence is the sole evidence presented, the Defence argument is thus essentially valid; however, it is not generally appropriate to use it as it stands when there is additional evidence in the case.

In a case of 'naked identification evidence' the story-line of the Defence argument may well be more appealing than a dry application of Bayes's Theorem, and would deliver the correct inference. However, one problem with reframing the Bayesian argument in this way is that, when the match probability is sufficiently tiny, the 'expected number of innocent matches' can be much less than 1. If, for example, the match probability in the case above had been 1 in 1 billion (as is now fairly routine in DNA profiling), rather than 1 in 10 million, we would expect only 0.03 innocent matches in a population of size 30 million. Applied formally, the Defence argument still applies: of those matching, there is one who is guilty and 0.03 who are innocent, and so the posterior probability of guilt is 1/1.03 = 0.971, exactly as would emerge from the Bayesian calculation. But when it requires them to juggle with small fractions of a person, the intuitive appeal of the Defence argument to the jury may well disappear.

We now turn to reconsider the Prosecution argument. This is often termed the 'Prosecutor's Fallacy', and indeed as presented it does turn on a serious mis-representation. This is not usually deliberate, but rather a consequence of the difficulty of expressing statements of conditional probability clearly and unambiguously in English, which sometimes seems to have been deliberately constructed to facilitate probabilistic misunderstanding (and I am sure that English is no different from any other natural language in this respect).

The match probability of 1 in 10 million is a measure of the probability, $P(M|\bar{G})$, of obtaining the match, on the hypothesis that the suspect is inno-cent. (Note that this is just the kind of conditional probability, for some observed event conditional on one or more entertained hypotheses, which underlies the Neyman–Pearson logic criticized in § 2.2 above.) However, the Prosecutor misinterprets this as $P(\bar{G}|M)$, the probability that the suspect is innocent, on the evidence of the match. This common and seductive error is also known as 'transposing the conditional'. In general, there is absolutely no reason for the two conditional probabilities, $P(M|\bar{G})$ and $P(\bar{G}|M)$, to be simi-lar. The actual connection between them is governed by Bayes's Theorem, which also requires other input, namely the prior probability of guilt, $P(G)$.

It is not hard to show that the transposition involved in the Prosecution argument *will* lead to an (approximately) valid result only in the special case that $P(G|B) = 1/2$ — i.e. when, on the basis of other evidence in the case, it can be regarded as equally probable that the suspect is or is not guilty. There may occasionally be cases where this is a reasonable assumption, but it is far from being so automatically. Contrast it with the implicit assumption under-lying the Defence argument — that the prior probability that the suspect is guilty is 1 in 30 million (or whatever the appropriate value for the population of possible perpetrators may be).

In summary, seen through Bayesian spectacles, both the Prosecution and the Defence arguments are generally inappropriate. Each becomes reasonable under certain specific assumptions about the prior probability of guilt, but these implicit assumptions are radically different in the two cases — which in turn explains why their answers are so completely different. Only the complete Bayesian argument is flexible enough to allow incorporation of whatever might be a reasonable prior judgment as to guilt, based on the other background evidence in the case.

3.4 Regina v. Denis John Adams

The case of Denis John Adams illustrates both the Bayesian approach to reasoning about identification evidence in the presence of other evidence, and some of the pitfalls besetting presentation of this approach in court.

Adams was arrested for rape. The only evidence linking him to the crime, other than the fact that he lived in the local area, was a match between his DNA and that of semen obtained from the victim. The relevant match probability was said to be 1 in 200 million, although the Defence challenged this, suggesting that a figure of 1 in 20 million or even 1 in 2 million could be more appropriate.

All other evidence in the case was in favour of the defendant. The victim did not pick Adams out at an identification parade, and had said that he did not resemble her attacker. Adams was 37 and looked older; the victim claimed the rapist was in his early twenties. Furthermore, Adams's girlfriend testified that he had spent the night of the attack with her, and this alibi remained unchallenged.

At trial, with the consent of both sides and the court, the jury was given instruction in the correct (Bayesian) way to combine all the evidence, introducing in turn (i) the prior probability, ahead of all specific evidence; (ii) likelihood ratios engendered by the defence evidence; and finally (iii) the identification evidence. Attention was drawn to the relevant probability questions to address, and the jurors were asked to assess their own probability values for these. The jury was guided through a practice calculation; however, care was taken at all times not to propose any specific probability values (other than match probabilities) to the jury. Below I shall insert specific numbers, but these are entirely hypothetical, and used merely to illustrate the general shape of the argument as it might be conducted by a juror following the instructions given.

It was indicated that there were approximately 150,000 males aged between 18 and 60 in the local area who, absent any other evidence, might have committed the crime. In order to allow for some possibility (assessed at something like a 25 per cent chance) that the attacker came from outside the area, a prior probability of guilt of the order of 1 in 200,000 might be considered reasonable. (The prior odds would then be essentially the same.)

With regard to the evidence that the victim did not recognize Adams as her attacker, one might assess the conditional probability of this happening, if Adams were truly guilty, at around 10 per cent; and its conditional probability, were he innocent, at around 90 per cent. On forming the required ratio of these figures, a likelihood ratio of 1/9 is obtained. As for the alibi evidence, one could assess this might be proffered with probability 25 per cent if he were guilty, as against 50 per cent if innocent, leading to a likelihood ratio of 1/2. If we assume that the two items of defence evidence would arise independently, given either guilt or innocence, then we can multiply them together to obtain an overall 'defence likelihood ratio' of 1/18.

Applying Bayes's Theorem to combine the prior odds of 1 in 200,000 and the above likelihood ratio of 1/18, the odds in favour of guilt, after taking into account the defence evidence but before incorporating the DNA evidence, would be assessed at 1 in 3.6 million.

Now the DNA match evidence by itself provides a likelihood ratio of between 200 million and 2 million in favour of guilt. Overwhelming though this may seem, it has to be taken in conjunction with the counterbalancing Defence evidence, by applying it (using Bayes's Theorem) to the previously calculated odds of 1 in 3.6 million. When this is done, we find that the posterior probability of guilt, given the totality of the evidence, varies from 0.98 (using a match probability of 1 in 200 million) to 0.36 (using 1 in 2 million). The Defence argued that, in the light of all the evidence, Adams's guilt had not been established beyond a reasonable doubt.

We cannot know what went on in the jury room, but the jury returned a verdict of guilty. The case then went to appeal. The Appeal Court roundly rejected the attempt to school the jury in the rational analysis of probabilistic evidence, saying that 'it trespasses on an area peculiarly and exclusively within the province of the jury', and that 'to introduce Bayes's Theorem, or any similar method, into a criminal trial plunges the jury into inappropriate and unnecessary realms of theory and complexity'. The task of the jury was said to be to 'evaluate evidence and reach a conclusion not by means of a formula, mathematical or otherwise, but by the joint application of their individual common sense and knowledge of the world to the evidence before them'. While one may well applaud this restatement of the traditional role of the jury, it fails to address the problem that common sense usually fares extremely badly when it comes to manipulating probabilities, and in particular common experience simply does not encompass such tiny probabilities as arise with DNA match evidence.

The appeal was granted on the basis that the trial judge had not adequately dealt with the question of what the jury should do if they did not want to use Bayes's Theorem. A retrial was ordered. Once again attempts were made to describe the Bayesian approach to the integration of all the evidence, once again the jury convicted, once again the case went to appeal and once again the Bayesian approach was rejected as inappropriate to the courtroom — although this time the appeal was dismissed.

At the time of writing, it seems that Bayesian arguments, while not formally banned from court, need to be presented with a good deal of circumspection. One possible approach (recommended by the Court of Appeal in the case of R. v. Doheny) uses a variation on the Defence argument presented in § 3.1.2. In that argument, the defendant was identified as one of four matching individuals, just one of whom is guilty. Absent other evidence, we can regard all four as equally likely to be the guilty party, leading to a probability of 1 in 4 that it is the defendant. However, we could go on to take into account other specific evidence, for or against the defendant, so generating differing probabilities of guilt for these four. This could be done formally, using Bayes's Theorem to combine a suitable likelihood ratio based on the other evidence with a 'prior' odds of 1/3, or informally by the application of 'common sense'. Thus in the Adams case (taking a base population of 200,000 and a match probability of 1 in 2 million), the number of innocent individuals matching the DNA would be expected to be 1/10. Taking into account the likelihood ratio of 1/18 from the other evidence means we must count Adams himself as 1/18 of an individual. The probability that it is Adams, rather than anyone else, who is the guilty party is thus calculated as $(1/18)/\{(1/18) + (1/10)\} = 0.36$, as before. But it must be admitted that such juggling with parts of people might be distasteful to the jury!

There is an additional wrinkle to the Adams story. At appeal, the Defence pointed out that Denis John Adams had a full brother, whose DNA had not been investigated. The probability that his brother had the same DNA profile as he did was calculated as 1 in 220, and it was submitted that this weakened the impact of DNA evidence against Denis John Adams. The Appeal Court dismissed this point on the grounds that there was no evidence that the brother might have committed the offence — ignoring the fact that, in the absence of the DNA match, neither was there any such evidence against Denis John Adams. The Bayesian argument, in its original or variant forms, does allow one to account for different match probabilities due to genetic relatedness. In effect, the brother adds an additional 1/120 of a person to the pool (previously 1/10) of matching individuals. This makes little numerical difference to the calculation as conducted above, but would have a much larger effect if the match probability for unrelated individuals had been taken as 1 in 200 million, so yielding 1/1000 as the number of unrelated matching individuals. Also, taking into account the existence of a number of individuals (unknown as well as known) related to Denis John Adams to various degrees could further add to the pool, perhaps boosting the figure of 1/120 substantially.

4. Databases and search

The police now have computer databases of tens of thousands of DNA profiles, and these look set to get much larger still. Increasingly, in crimes such as rape or murder where there is no obvious suspect, such a database is 'trawled' to see if it contains a profile matching that found at the scene. If it does, the matching individual may be arrested and charged, even in the absence of any other evidence linking him to the crime.

Because a jury might be unfairly prejudiced by the information that a defendant's profile had previously been entered into a police database, the fact that he was identified in this way would typically not be admissible as evidence in court. Nevertheless, for the sake of rational analysis, it is reasonable to ask how — if at all — the fact of the database search should affect the impact of the DNA match. Once again we can identify two mutually contradictory lines of argument.

4.1 The Prosecution line

Prosecuting Counsel argues:

> A database of 10,000 profiles has been searched, and only the defendant's profile has been found to match that found at the scene. We have thus eliminated 9,999 potential alternative suspects. This must make it more likely that the defendant is guilty.

According to this argument, the evidence for the defendant's guilt is stronger (although, given the initial extremely large number of potential suspects, perhaps only marginally) than it would have been had he been identified without searching the database. It also gives some comfort that the usual practice of hiding the database trawl from the jury is of little consequence, and is if anything conservative, erring on the side of the defendant.

4.2 The Defence line

Counsel for the Defence argues:

> The DNA profile found at the scene of the crime occurs in the population with a frequency of 1 in 1 million. The police database contains 10,000 profiles. The probability that, on searching the database, a match will be found is thus $10,000 \times 1/1,000,000 = 1/100$. This figure, rather than 1 in 1 million, is the relevant match probability. Clearly it is not nearly small enough to be regarded as convincing evidence against my client.

From this point of view, the effect of the database search is to weaken, very dramatically, the strength of the evidence against the defendant.

4.3 Disagreeing to disagree

The comparative merits of the above two arguments have been fiercely debated in the statistical community, with those of a Neyman–Pearson frame of mind favouring the Defence line (National Research Council, 1996; Stockmarr, 1999), and Bayesians arguing in favour of the Prosecution line, which indeed can be rephrased as a Bayesian probability calculation (Dawid and Mortera, 1996; Balding and Donnelly, 1996; Donnelly and Friedman, 1999; Dawid, 2001). Rarely can there have been such an important application of statistics in which the differing intuitions and approaches of the two schools lead to answers so vividly and violently opposed. Defence Counsel, and Neyman–Pearsonites, claim that the upward adjustment of the match probability is essential, to allow for the fact that the hypothesis under test (that the specific matching individual identified by the search did indeed commit the crime) was not known in advance, but only formulated in the light of the data examined. Prosecuting Counsel, and Bayesians, counterclaim that this is irrelevant, and that it is the singular guilt of the actual defendant that is at issue, not the whole database that is on trial. One argument that seems particularly telling to me is to consider what happens when we take matters to extremes. Thus suppose that (as may indeed soon become the case) the police database is fully comprehensive, containing the DNA profiles of all members of the population. And suppose that exactly one of these is found to match the DNA profile from the scene of the crime. On purely logical (and common-sense) grounds, we then know for sure that we must have identified the true culprit — and this is exactly the import of the Prosecution line. But similarly to take the Defence line to extremes would be to regard the evidence against the defendant as weakened to the point of non-existence, which is clearly absurd.

Statisticians on both sides of this divide (and here I speak as one firmly planted on the Prosecution/Bayesian side) can only despair of the inability of those on the other to appreciate the strengths of the 'correct' arguments, and the weaknesses of the 'incorrect' arguments. An important moral is that, while all agree that arguments phrased in terms of probabilities are essential to such rational argument, there are various different ways of formulating such arguments — and it does matter how this is done. I personally find that looking at the world through Bayesian spectacles gives a crystal-clear image, and brings into focus many issues that would otherwise remain fuzzy and confused.

5. Concluding remarks

The last message is the one that I would like the reader to take home. Bayesian statistics is just the logic of rational inference in the presence of uncertainty. It is a valuable intellectual resource, bringing clarity to the formulation and analysis of many perplexing problems. I believe that it could be of far greater significance in the law than has thus far been allowed or appreciated.

I conclude by briefly mentioning some other problems of legal reasoning that have been greatly clarified by examining them from a Bayesian perspective.

5.1 Genetic heterogeneity

There has been considerable technical discussion about how to allow, for purposes of DNA identification, for the fact that the genetic structure of the population is heterogeneous, so that a given DNA profile may generate different 'match probabilities', depending on the subpopulation that is used to determine the frequencies of its constituent alleles (National Research Council, 1992, Chapter 3). And again there has been disagreement and confusion about appropriate ways to handle this problem (Foreman et al., 1997; Roeder et al., 1998). A fully Bayesian approach (Dawid, 1998; Dawid and Pueschel, 1999) reveals that at least some of the seeming disagreements are merely semantic, while identifying other inadequacies in the arguments presented.

5.2 Combining evidence

The Bayesian machinery is ideally suited to the modelling and analysis of complex interrelations between many variables. Recent years have seen the development of powerful computational systems based on natural graphical representations (Cowell et al., 1999) — an enterprise that can be seen as developing on the pioneering work of Wigmore (1931) (see in particular Kadane and Schum, 1996). Dawid and Evett (1997) show how these ideas can be used to organize and implement the combination of evidence from a variety of sources, such as forensic evidence obtained from fibre analysis and from bloodstains.

5.3 Missing data

Particularly challenging problems arise when DNA evidence is not available on the principal suspect, but indirect evidence can be obtained by typing close relatives. This arises quite commonly in paternity suits, and occasionally in criminal cases, but hitherto there has been no clear understanding of how to analyse such problems. By building a suitable graphical computer model, it is possible readily to calculate the appropriate likelihood ratio, taking correct account of the information that is actually available (Dawid et al., 2001).

Note. I am grateful to Peter Donnelly and Julia Mortera for comments on an earlier draft of this paper.

References

Aitken, C. G. G. (1995), *Statistics and the Evaluation of Evidence for Forensic Scientists*, Chichester: John Wiley and Sons.

Anderson, T. and Twining, W. (1991), *Analysis of Evidence*, London: Weidenfeld and Nicolson.

Balding, D. J. and Donnelly, P. J. (1996), 'DNA profile evidence when the suspect is identified through a database search', *Journal of Forensic Sciences* **41**: 603–7.

Cowell, R. G., Dawid, A. P., Lauritzen, S., and Spiegelhalter, D. J. (1999), *Probabilistic Networks and Expert Systems*, New York: Springer.

Dawid, A. P. (1987), 'The difficulty about conjunction', *The Statistician* **36**: 91–7.

Dawid, A. P. (1994), 'The island problem: Coherent use of identification evidence', in *Aspects of Uncertainty: A Tribute to D. V. Lindley* (ed. P. R. Freeman and A. F. M. Smith), ch. 11, pp. 159–70, Chichester: John Wiley and Sons.

Dawid, A. P. (1998), 'Modelling issues in forensic inference', in *1997 ASA Proceedings, Section on Bayesian Statistics*, pp. 182–6.

Dawid, A. P. (2001), 'Comment on Stockmarr's "Likelihood ratios for evaluating DNA evidence when the suspect is found through a database search"' (with response by Stockmarr), *Biometrics* **57**: 234–7.

Dawid, A. P. and Evett, I. W. (1997), 'Using a graphical method to assist the evaluation of complicated patterns of evidence', *Journal of Forensic Sciences* **42**: 226–31.

Dawid, A. P. and Mortera, J. (1996), 'Coherent analysis of forensic identification evidence', *Journal of the Royal Statistical Society, Series B* **58**: 425–43.

Dawid, A. P. and Mortera, J. (1998), 'Forensic identification with imperfect evidence', *Biometrika* **85**: 835–49.

Dawid, A. P. and Pueschel, J. (1999), 'Hierarchical models for DNA profiling using heterogeneous databases', in *Bayesian Statistics 6* (ed. J. M. Bernardo, J. O. Berger, A. P. Dawid and A. F. M. Smith). Oxford: Oxford University Press, pp. 187–212.

Dawid, A. P., Mortera, J., Pascali, V. L., and van Boxel, D. W. (2001), 'Probabilistic expert systems for forensic inference from genetic markers', *Scandinavian Journal of Statistics*, in press.

Donnelly, P. J. and Friedman, R. D. (1999). 'DNA database searches and the legal consumption of scientific evidence', *Michigan Law Review* **97**: 931–84.

Eggleston, R. (1983), *Evidence, Proof and Probability*, 2nd edn, London: Weidenfeld and Nicolson.

Evett, I. W. and Weir, B. S. (1998), *Interpreting DNA Evidence*, Sunderland, MA: Sinauer.

Foreman, L. A., Smith, A. F. M., and Evett, I. W. (1997), 'Bayesian analysis of deoxyribonucleic acid profiling data in forensic identification applications (with Discussion)', *Journal of the Royal Statistical Society, Series A* **160**: 429–69.

Gastwirth, J. L. (ed.) (2000), *Statistical Science in the Courtroom*, New York: Springer-Verlag.

Kadane, J. B. and Schum, D. A. (1996), *A Probabilistic Analysis of the Sacco and Vanzetti Evidence*, New York: John Wiley and Sons.

National Research Council (1992), *DNA Technology in Forensic Science*, Washington, DC: National Academy Press.

National Research Council (1996), *The Evaluation of Forensic DNA Evidence*, Washington, DC: National Academy Press.

Robertson, B. and Vignaux, G. A. (1995), *Interpreting Evidence*, Chichester: John Wiley and Sons.

Roeder, K., Escobar, M., Kadane, J. B., and Balazs, I. (1998). 'Measuring heterogeneity in forensic databases using hierarchical Bayes models', *Biometrika* **85**: 269–87.

Stockmarr, A. (1999), 'Likelihood ratios for evaluating DNA evidence when the suspect is found through a database search', *Biometrics* **55**: 671–7.

Wigmore, J. H. (1931), *The Science of Judicial Proof*, 2nd edn, Boston: Little, Brown.

5

Bayes, Hume, Price, and Miracles

JOHN EARMAN

MY TOPIC is the Bayesian analysis of miracles. To make the discussion concrete, I will set it in the context of eighteenth-century debate on miracles, and I will focus on the response to David Hume's celebrated argument against miracles that Thomas Bayes would have made and did in part make, albeit from beyond the grave, through his colleague Richard Price.

1. My trinity: Bayes, Price, and Hume

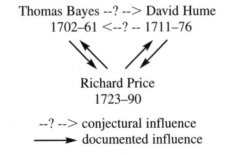

Thomas Bayes --? --> David Hume
1702–61 <--? -- 1711–76

Richard Price
1723–90

--? --> conjectural influence
———► documented influence

It is irresistible to think that Bayes had read Hume and that Bayes's 'Essay Towards Solving a Problem in the Doctrine of Chances' was at least in part a reaction to Hume's sceptical attack on induction.[1] But while the Appendix of the published paper is surely making reference to Hume, that Appendix was penned not by Bayes but by Richard Price. It is also nice to think that Hume read Bayes's essay and then to speculate about what Hume's reaction would have been. But while there is a reference to Bayes's essay in Price's *Four Dissertations*, a copy of which was sent to Hume who duly acknowledged receiving and reading it (see below), there is no evidence that Hume followed up the reference. And even if he had, it is unlikely that he would have

---

[1] Bayes's essay was published posthumously in 1763. It was probably written in the 1740s.

understood Bayes's essay since he was largely innocent of the technical developments that were taking place in the probability calculus.

While there is no evidence of a direct connection between Bayes and Hume, the indirect connection that goes through Price is solid. Although we know little of the relationship between Bayes and Price, it must have been reasonably close since Bayes's will left £200 to be divided between Price and one John Boyl (see Barnard, 1958) and since it was Price who arranged for the posthumous publication of Bayes's essay. In the other direction, Price was a persistent critic of Hume — not just on induction but on matters of religion and ethics as well. Despite the sharp differences in opinion the two men remained on remarkably cordial terms, dining together in London and in Price's home in Newington Green (see Thomas, 1924). In the second edition of *Four Dissertations* Hume is lauded by Price as 'a writer whose genius and abilities are so distinguished as to be above any of my commendations' (1768, p. 382). And Hume in turn praised Price for the 'civility which you have treated me' (Klibansky and Mossner, 1954, 233–4). More intriguingly, Hume goes on to say that 'I own to you, that the Light in which you have put this Controversy [about miracles] is new and plausible and ingenious, and perhaps solid'. Unfortunately, Hume left the matter hanging by adding that 'I must have some more time to weigh it, before I can pronounce this Judgment with satisfaction to myself' (ibid.).

I will finish Hume's unfinished task. I will claim that Price's criticisms of Hume's argument against miracles were largely solid. More generally, I claim that when Hume's 'Of miracles' is examined through the lens of Bayesianism, it is seen to be a shambles.[2]

2. The context

The debate on miracles which took place in eighteenth-century Britain is rich and endlessly fascinating. One of the key philosophical problems underlying the debate was posed before the beginning of the century in Locke's *Essay Concerning Human Understanding* (1690); namely, how should one apportion belief when the two main sources of credibility, 'common observation in like cases' and 'particular testimonies', are at odds? The case of eyewitness

---

[2] A first draft of Hume's essay on miracles was written (probably) in 1737. But the essay did not appear in print until 1748 when it was published as Chapter 10 ('Of miracles') of Hume's *Philosophical Essays Concerning Human Understanding*, later called *Enquiries Concerning Human Understanding*. The internal evidence indicates that the published version of Hume's miracles essay was written in the 1740s.

testimony to miraculous events — the central focus of Hume's 'Of miracles' — presents the extreme form of this problem. The miraculous event around which the eighteenth-century debate swirled was, of course, the Resurrection of Jesus of Nazareth, though the more prudent naysayers took care to make oblique reference to this matter — Hume, for example, couches his discussion in terms of the hypothetical case of the resurrection of Queen Elizabeth.

To convey a bit of the flavour of the debate I will trace one of the many threads — the one that starts with Thomas Woolston, one of the incautious naysayers. His *Six Discourses on the Miracles of Our Savior* (1727–29) was an undisguised and broadsided attack on the New Testament miracles. As attested by Swift, it created a minor sensation:[3]

> Here is Woolston's tract, the twelfth edition
> 'Tis read by every politician:
> The country members when in town
> To all their boroughs send them down:
> You never met a thing so smart;
> The courtiers have them all by heart.

What the courtiers learned from reading Woolston was that the New Testament accounts of miracles were filled with 'absurdities, improbabilities, and incredibilities'. While Woolston's charges and his sarcasm and obvious contempt for the Church authorities won him a large readership — reportedly, 30,000 copies of *Six Discourses* were printed — they also earned him a fine and a stay in prison, where he died.

Six Discourses received a number of replies, the most influential being Thomas Sherlock's *Tryal of the Witnesses* (1728), which itself went through fourteen editions. Sherlock was answered by Peter Annet (1744a,b), and Annet in turn was answered by Chandler (1744), Jackson (1744), and West (1747). And so it went.

Set in this context, Hume's essay on miracles seems both tame and derivative. It is also something of a muddle. What appears on first reading to be a powerful and seamless argument turns on closer examination to be series of considerations that don't comfortably mesh. And worse, Hume's thesis remains obscure — the possibilities range from the banal to the absurd:

(T1) One should be cautious in accepting testimony to marvelous and miraculous events, and doubly so when religious themes arise.

(T2) The testimonies of the Disciples do not establish the credibility of the resurrection of Jesus.

---

[3] Quoted in Burns (1981). Burns's book provides a good overview of the eighteenth-century miracles debate.

(T3) In no recorded case does the testimony of eyewitnesses establish the credibility of a miracle deemed to have religious significance.

(T4) Eyewitness testimony is incapable of establishing the credibility of a miracle deemed to have religious significance.

(T5) Eyewitness testimony is incapable of establishing the credibility of any miraculous event.

Hume deserves no credit — save for pompous solemnity — for uttering various forms of the banality (T1), since even the pro-miracle proponents in the eighteenth-century debate were at pains to acknowledge the pitfalls of eyewitness testimony and the need to carefully sift the evidence, especially in the case of religious miracles. Hume believed (T2), but refused to join Annet, Sherlock, Woolston et al. in arguing the specifics. He surely also believed something in the neighbourhood of (T3), but his cursory recitation and rejection in Part II of his essay of a number of sacred and profane miracles hardly counts as much of an argument for this sweeping thesis. What Hume manages beautifully is the creation of an illusion that he was in possession of principles for evaluating evidence that allowed him to remain above the nastiness of the fray. This would have been the case if he had been in possession of principles that entailed (T4) or (T5). But there are no such principles since (T4) and (T5) are absurd. A skilful oscillation among these various theses is what helps to create the illusion of a worthy argument against miracles.

Stripped of the theology, the key issue addressed in Hume's essay is how to evaluate the evidential force of fallible witnesses. The obvious tool to apply was the probability calculus, and one of the first published applications in English was the anonymous (George Hooper?) essay 'A Calculation of the Credibility of Human Testimony', which appeared in the 1699 *Transactions of the Royal Society*. Some advances were made in the eighteenth century — for example, by Price — but it was not until the work of Laplace (1812, 1814) and Charles Babbage (1838) that definitive results were obtained. Hume's pronouncements on the key issue have that combination of vagueness, obscurity, and aphorism that philosophers find irresistible, and, consequently, have led to an endless and largely unfruitful debate in the philosophical literature, the contributors to which are willing to go to extraordinary lengths in an attempt to show that Hume had something interesting to say on the matter.

To state what should be obvious, but still manages to elude some commentators, the core issues in Hume's essay are independent of the theological subtleties of that vexed term 'miracle'. For what an eyewitness testifies to in the first instance is the occurrence of an event which can be characterized in purely naturalistic terms, e.g., the return to life of a dead man. How such an event, if its credibility is established by eyewitness testimony, can serve theological purposes is a matter I will take up in due course. But first I have to substantiate my negative evaluation of Hume's treatment of eyewitness testimony. At the same time I hope to show how the issues can be advanced with the help of Bayes's apparatus.

3. Hume's 'proof' against miracles

Here is Hume's 'proof', such at it is:

> A miracle is a violation of the laws of nature; and as a firm and unalterable experience has established these laws, the proof against a miracle, from the very nature of the fact, is as entire as any argument from experience can possibly be imagined. (1748, p. 114)

For Hume 'proofs' are 'such arguments from experience as leave no room for doubt or opposition' (1748, p. 56, fn. 1). He thought his proof against miracles fitted this bill since it is 'full and certain when taken alone, because it implies no doubt, as in the case of all probabilities' (Greig, 1932, Vol. I, p. 350). In probability language, Hume seems to be saying that if L is a lawlike generalization, such as 'No dead man returns to life', and E records uniform past experience in favour of L, then $P(L|E) = 1$; hence, if M asserts a violation of L, then it follows from the rules of probability that $P(M|E) = 0$.

Price opposed Hume's wildly incautious and inaccurate account of inductive practice: 'It must be remembered, that the greatest uniformity and frequency of experience will not offer a proper proof, that an event will happen in a future trial.' But he continued with a streak of dark pessimism: 'or even render it so much as probable that it will always happen in all future trials' (1768, pp. 392–3). Here Price is reporting a result that follows from Bayes's interesting but ultimately flawed attempt to justify a particular prior probability assignment that had the consequences that, in the case of a domain with a countably infinite number of individuals, (i) $P(L|E) = 0$ no matter how extensive the past record E of uniform experience in favour of L is, and (ii) nevertheless, the probability that the next instance conforms to L approaches 1 as the number of past positive instances increases without bound. Leaving aside the issue of the justification of prior probabilities, the relevant thing to note for present purposes is that Hume's account of inductive practice is descriptively inaccurate: scientists don't think that uniform experience in favour of a lawlike generalization leaves no room for doubt — if they did it would be hard to explain their continued efforts to search for exceptions.

But the crucial point is that Hume recognized that his 'proof' only applies when the evidence consists of uniform experience (recall the qualifier 'when taken alone'). The real issue is joined when that proof is opposed by a counterproof from eyewitness testimony. Many commentators from Hume's day to the present have read Hume as saying that this contest has a preordained outcome. The idea is that the probative force of eyewitness testimony derives from the observation of the conformity between the testimony and reality; but whereas our experience in favour of the relevant lawlike generalization and, thus, against an exception that constitutes a miracle, is uniform, our experience as to the trustworthiness of witnesses is anything but uniform. Thus, Price took Hume to be arguing that to believe a miracle on the basis of testimony is to 'prefer a weaker proof to a stronger' (1768, p. 385).[4] Such a reading makes Hume's essay into a puzzle: if this was Hume's argument, why did the essay have to be more than one page long? And apart from the puzzle, the position being attributed to Hume implies the absurd thesis (T5).

The anti-miracle forces in the eighteenth-century debate sometimes asserted that uniform experience always trumps testimony. Such assertions were met with variations of Locke's example of the king of Siam who had never seen water and refused to believe the Dutch ambassador's report that, during the winter in Holland, water became so hard as to support the weight of an elephant. Thus, in the *Tryal of the Witnesses* (1728), published almost a decade before Hume got the idea for his miracle essay,[5] Sherlock puts the rhetorical question: '[W]hen the Thing testify'd is contrary to the Order of Nature, and, at first sight at least, impossible, what Evidence can be sufficient to overturn the constant Evidence of Nature, which she gives us in the constant and regular Method of her Operation?' (1728, p. 58). Sherlock answers his rhetorical question with a version of Locke's example: what do the naysayers against the Resurrection have to say 'more than any man who never saw Ice might say against an hundred honest witnesses, who assert that Water turns into Ice in cold Climates?' (1728, p. 60). That Hume's contemporaries were not impressed by his miracles essay is partly explained by the fact that he seemed to be offering a warmed-up version of the past debate, and

---

[4] And C. D. Broad took Hume to be saying that 'we have never the right to believe any alleged miracle however strong the testimony for it may be' (1916–17, p. 80).

[5] Hume tell us that he got the idea during his stay in La Flèche (1736–37); see Greig (1932), Vol. I, p. 361.

a defective version at that, since the 1748 edition of the *Enquiries* is silent on the issues raised by Locke's example.[6]

Let's be charitable to Hume by not subscribing to the reading that takes him to be saying that uniform experience always trumps eyewitness testimony. The issue then becomes how to tell when the balance tips in favour of one or the other. Hume's famous 'maxim' might be thought to provide just such a prescription.

4. Hume's maxim

Hume announces the 'general maxim'

> That no testimony is sufficient to establish a miracle, unless the testimony be of such a kind, that its falsehood would be more miraculous, than the fact, which it endeavours to establish. (1748, pp. 115–16)

Most commentators have seen profound wisdom here. I see only triviality.

Suppose that we are in a situation where witnesses have offered testimony $t(M)$ to the occurrence of the miraculous event M. Let E be the record of our past experience in favour of the lawlike generalization to which M is an exception. Then, as good Bayesians, our current degree of belief function should be the conditionalization on $t(M)\&E$ of the function $P(\cdot)$ we had before obtaining this evidence. Thus, the relevant probability of the event which the testimony endeavours to establish is $P(M|t(M)\&E)$, while the relevant probability of the falsehood of the testimony is $P(\sim M|t(M)\&E)$. To say that the falsehood of the testimony is more miraculous than the event which it endeavours to establish is just to say that the latter probability is smaller than the former, i.e.

$$P(M|t(M)\&E) > P(\sim M|t(M)\&E) \tag{1}$$

which is equivalent to

$$P(M|t(M)\&E) > 0.5. \tag{2}$$

On this reading, Hume's maxim is the correct but unhelpful principle that no testimony is sufficient to establish the credibility of a miracle unless the testimony makes the miracle more likely than not.

---

[6] Hume's 'Indian prince' example makes its appearance in a hastily added endnote to the 1750 edition. In later editions a paragraph and a long footnote on the Indian prince were added to the text.

A number of other renderings of Hume's maxim have been offered, but either they fail to do justice to the text of Hume's essay or else they turn Hume's maxim into a false principle (see Earman, 2000 for details). For example, Gillies (1991) and Sobel (1991) translate Hume's maxim as

$$P(M|E) > P(\sim M \& t(M)|E) \qquad (GS)$$

(*GS*) does provide a necessary condition for the credibility of a miracle in the sense of (1)–(2). But the context in which the maxim is enunciated makes it clear that Hume intended to provide a sufficient as well as a necessary condition for testimony to establish the credibility of a miracle.[7] In modern conditional probability notation Price's (1767) reading could be rendered in two ways by moving either the $\sim M$ or the $t(M)$ in (*GS*) to the right of the modulus:

$$P(M|E) > P(\sim M|t(M)\&E) \qquad (P)$$

or

$$P(M|E) > P(t(M)|\sim M\&E) \qquad (P')$$

Hume's text is ambiguous enough that either (*P*) or (*P′*) could, without forcing, serve as a translation. But each of these conditions can fail even when (1)–(2) hold.

In closing this section I want to acknowledge Colin Howson's (2000) insight that Hume's maxim takes on a different complexion if 'establish a miracle' means render certain rather than render credible. This reading, however, makes Hume's argument ineffective against his more sophisticated eighteenth-century opponents who took the aim of natural religion as establishing the likelihood rather than the certainty of religious doctrines. John Wilkins, a founder of the Royal Society and a spokesman for liberal Anglicans, held that

> 'Tis sufficient that matters of Faith and Religion be propounded in such a way, as to render them highly credible, so as an honest and teachable man may willingly and safely assent to them, and according to the rules of Prudence be justified in so doing. (1699, p. 30)

In a similar vein, Samuel Clarke wrote that

> [S]uch moral Evidence, or mixt Proofs from Circumstances and Testimony, as most Matters of Fact are only capable of, and wise and honest Men are always satisfied

---

[7] Hume writes: 'If the falsehood of his testimony would be more miraculous, than the event which he relates; *then, and not till then*, can he pretend to commend my belief or opinion' (p. 116). I take the italicized phrase to have the force of 'if and only if'.

with, ought to be accounted sufficient for the present Case [the truth of the Christian revelation]. (1705, p. 600)

And John Tillotson, on whom Hume bestows (ironic?) praise at the beginning of his miracles essay, wrote:

And for any man to urge that tho' men in temporal affairs proceed upon moral assurance, yet there is a greater assurance required to make men seek Heaven and avoid Hell, seems to me highly unreasonable. (1728)

5. Hume's diminution principle

This principle is enunciated by Hume in the following passage:

[T]he evidence, resulting from the testimony, admits of a diminution, greater or less, in proportion as the fact is more or less unusual. (1748, p. 113)

In citing the proverbial Roman saying, 'I should not believe such a story were it told me by Cato', Hume intimates that in the case of a miracle the diminution of the evidential force of testimony is total.

Price responded that 'improbabilities as such do not lessen the capacity of testimony to report the truth' (1768, p. 413). This is surely right, as was Price's further claim that the diminution effect operates through the factors of the intent to deceive and the danger of being deceived, either by others or by oneself. Unfortunately Price overstepped himself in claiming that when the first factor is absent, testimony 'communicates its own probability' to the event (1768, p. 414). It was left to Laplace (1814) to give a correct and thorough Bayesian analysis of when and how the intent to deceive and the danger of being deceived give rise to a diminution effect. Two cases suffice to illustrate the results that can emerge from the analysis.

The first case — one considered by Price — concerns a witness who testifies $t(W_{79})$ that the winning ticket in a fair lottery with N tickets is #79. Assuming that when the witness misreports, she has no tendency to report one wrong number over another, then

$$P(W_{79}|t(W_{79})\&E) = P(t(W_{79})|(W_{79})\&E) \tag{3}$$

where E records the background knowledge of the lottery. And this is so regardless of how large N is and, thus, regardless of how small the prior probability $P(W_{79}|E) = 1/N$. In such a case the testimony does, in Price's words, communicate its own probability to the event. But, as Price was well aware, (3) does not hold if the witness has some ulterior interest in reporting #79 as the winner.

The second case concerns the testimony $t(W)$ that a ball drawn at random from an urn containing one white ball and $N-1$ black balls is white. As long as there is a non-zero probability that the witness misperceives the colour of the ball drawn, or else there is a non-zero probability that the witness misreports the correctly perceived colour, the posterior probability $P(W|t(W)\&E)$ does diminish as N is increased and, thus, as the prior probability $P(W|E) = 1/(N-1)$ is reduced.

The difference between the two cases lies in the fact that in the urn case, as the prior probability is reduced the likelihood factor $P(t(W)|\neg W\&E)|$ $P(t(W)|W\&E)$ remains the same because the visual stimulus presented to the witness when either a black or a white ball is drawn is independent of the numbers of black and white balls in the urn. By contrast, in the lottery case the corresponding likelihood factor changes in such a way as to cancel the change in the prior factor, quenching the diminution effect.

Two comments are in order. First, I see no a priori reason to think that cases of reported miracles are always or even mostly like the urn case in that they are subject to the diminution effect. Even when the reported miracle is subject to the diminution effect, Hume gets his intended moral only for one order of quantifiers; namely, for any given fallible witness (who may practise deceit or who is subject to misperception, deceit, or self-deception), there is a story so a priori implausible that we should not believe the story if told by that witness. It does not follow that there is a story so a priori implausible (unless, of course, the prior probability is flatly zero) that for any fallible witness, we should not believe the story if told by that witness. To get this latter implication one needs the extra postulate that there are in-principle bounds below which the probability of errors of reporting cannot be reduced. Hume's cynicism suggests such a postulate, but cynicism is not an argument.

6. Multiple witnesses

Hume makes some nods to the importance of multiple witnessing, but he seems not to have been aware of how powerful a consideration it can be. In fairness, the power was clearly and fully revealed only in the work of Laplace (1812, 1814) and more especially the work of Babbage (1838), whose *Ninth Bridgewater Treatise* devotes a chapter to a refutation of Hume's argument against miracles.

Suppose that there are N fallible witnesses, all of whom testify to the occurrence of an event M. For simplicity suppose that the witnesses are all equally fallible in that for all $i = 1, 2, ..., N$

$$P(t_i(M)|M\&E) = p, \qquad P(t_i(M)|\sim M\&E) = q \tag{4}$$

And suppose that conditional on the occurrence (or non-occurrence) of M, the testimonies of the witnesses are independent in the sense that

$$
\begin{aligned}
&P(t_1(M)\&t_2(M)\& \cdots \&t_N(M)| \pm M\&E) \\
&= P(t_1(M)| \pm M\&E)xP(t_2(M)| \pm M\&E)x \cdots xP(t_N(M)| \pm M\&E)
\end{aligned} \tag{5}
$$

where $\pm M$ stands for M or $\sim M$, and where the choice is made uniformly on both sides of the equality. Then the concurrent testimony of all N witnesses gives a posterior probability

$$P(M|t_1(M)\&t_2(M)\& \cdots \&t_N(M)\&E) = \cfrac{1}{1 + \left[\dfrac{P(\sim M|E)}{P(M|E)} \right]\left(\dfrac{q}{p} \right)^N} \tag{6}$$

The witnesses may be very fallible in the sense that q can be as close to 1 as you like. But as long as they are minimally reliable in the sense that $p > q$, it follows from (6) that the posterior probability of M can be pushed as close to 1 as you like by a sufficiently large cloud of such fallible but minimally reliable witnesses — provided, of course, that $P(M|E) > 0$. Using this result Babbage was able to give some nice examples where just twelve minimally reliable witnesses can push the posterior probability of an initially very improbable event to a respectably high level.

In sum, if we set aside Hume's wildly optimistic account of induction on which $P(M|E) = 0$, then he must agree that fallible multiple witnesses can establish the credibility of a miracle, provided that their testimonies are independent in the sense of (5) and that they are minimally reliable in the sense that $p > q$. Which of these provisos would Hume have rejected? The answer is 'Neither', at least in some cases of secular miracles, such as in his hypothetical example of eight days of total darkness around the world.

> [S]uppose, all authors, in all languages, agree, that, from the first of January 1600, there was total darkness over the whole earth for eight days: Suppose that the tradition of this extraordinary event is still strong and lively among the people: that all travellers, who return from foreign countries, bring us accounts of the same tradition without the least variation or contradiction: It is evident, that our present philosophers, instead of doubting the fact, ought to receive it as certain ... (1748, pp. 127–8)

But in cases to which religious significance is attached, Hume professed to be unswayed by even the largest cloud of witnesses. In the hypothetical case of the resurrection of Queen Elizabeth, Hume declared that he would 'not have the least inclination to believe so miraculous an event' (1748, p. 128), even if all the members of the court and Parliament proclaimed it. Here it is less plausible than in the previous case that the independence assumption is satisfied since the witnesses can be influenced by each other and the general hubbub surrounding the events. But there seems to be no in-principle difficulty in arranging the circumstances so as to secure the independence condition. The minimal reliability condition becomes suspect if the alleged resurrection is invested with religious significance because witnesses in the grip of religious fervour tend to be more credulous and because they may give in to the temptation to practise deceit in order to win over the unconverted. But it is an insult to the quality of the eighteenth-century debate to think that the participants needed a sermonette on this topic from Hume. The pro-miracle proponents were acutely aware of the need to scrutinize the contextual factors that might give clues as to the reliability of the witnesses. The conclusions they drew from their scrutiny may have been mistaken, but if so, the mistakes did not flow from a failure to heed the empty solemnities of Hume's essay. Hume's treatment of the hypothetical Queen Elizabeth case drips with cynicism:

> [S]hould this miracle be ascribed to any new system of religion; men, in all ages, have been so much imposed on by ridiculous stories of that kind, that this very circumstance would be a full proof of cheat, and sufficient, with all men of sense, not only to make them reject the fact, but even reject it without farther examination. (1748, pp. 128–9)

The cynicism remains just cynicism unless it is backed by an argument showing that, in principle, the witnesses cannot be minimally reliable and independent when the alleged miracle is ascribed to system of religion. Such an argument is not to be found in Hume's 'Of miracles'.

7. Miracles as a just foundation for religion

Grant that nothing in principle blocks the use of eyewitness testimony to establish the credibility of a miracle — say, a resurrection — of supposed religious significance. It might still seem that Hume has safe ground to which to retreat. On his behalf one can argue that to serve as a 'just foundation for religion', the miracle must not only satisfy Hume's first definition of 'miracle' as a violation of a (putative) law of nature but must also satisfy the second definition, according to which a miracle is 'a transgression of a law of nature by a particular volition of the Deity, or by the interposition of some invisible agent' (1748, p. 115n). But (still continuing on Hume's behalf, now in the voice of John Stuart Mill), nothing can ever prove that the resurrection is miraculous in the sense of Hume's second definition because 'there is still another possible hypothesis, that of its being the result of some unknown natural cause; and this possibility cannot be so completely shut out as to leave no alternative but that of admitting the existence and intervention of a being superior to nature' (1843, p. 440).

This line is used over and again across the decades by commentators sympathetic to Hume. It is apt to strike the innocent reader as a powerful consideration, but when the context is filled in it is seen as unavailing against the more sophisticated eighteenth-century pro-miracles proponents. For, to repeat, the role these proponents saw for miracles was not that of a direct and full proof of the presence of God by marks of a supernatural intervention in human affairs. Rather, the miraculous figured in an argument whose goal was to render religious doctrines highly credible and, ideally, to give them the kind of moral assurance needed to render a jury verdict of beyond reasonable doubt. And to fulfil this role, miracles need not be conceived as supernatural interventions; they need only serve as probabilistic indicators of the truth of the religious doctrines.

To illustrate how miracles can serve to confirm religious doctrines, suppose that testimonial evidence $t(M)$ has incrementally confirmed M:

$$P(M|t(M)\&E) > P(M|E) \tag{7}$$

Suppose also that the testimony bears on the religious doctrine D only through M in that

$$P(D|\pm M\& \pm t(M)\&E) = P(D|\pm M\&E) \tag{8}$$

And suppose finally that

$$P(M|D\&E) > P(M|\sim D\&E) \tag{9}$$

Then it follows that

$$P(D|t(M)\&E) > P(D|E) \tag{10}$$

that is, $t(M)$ incrementally confirms D.

Condition (8) is surely unobjectionable. Condition (9) is the sense in which M serves as a probabilistic indicator of the truth of D, and to fulfil this role M need not be the result of a supernatural intervention that disrupts the order of nature. And condition (9) is seemingly easy to satisfy. For example, isn't it obvious that the miracle of the loaves is more likely on the assumption that Christian doctrine is true than on the assumption that it is false? In fact, it is not obvious. In general, whether or not (9) is satisfied depends on what alternatives are included in $\sim D$ and what their prior probabilities are.[8] Consider, for example, the case where a high prior is given to the possibility that there is a non-Christian deceiver God who actualizes a world containing events designed to mislead people into falsely believing Christian doctrine. And even when D is incrementally confirmed by $t(M)$, there is no assurance that evidence of other miracles will push the probability of D anywhere near that required for moral certainty — again, it depends on the available alternatives and their prior probabilities.

So the discouraging word is that Bayesianism does not pave an easy road for religion. But by the same token there is no obvious difference here with theoretical physics: whether and how much a physical theory T that postulates unobservable properties of unobservable elementary particles is confirmed by direct or testimonial evidence about streaks in cloud chambers depends on what alternatives to T are entertained and what priors these alternatives are assigned.

8. Parting company

Let us agree for present purposes that the attempts by Bayes and others to objectify priors do not succeed.[9] And, again for present purposes, let us agree that the subjectivist form of Bayesianism captures the logic of inductive reasoning. This form of Bayesianism imposes two and only two constraints on degrees of belief: a synchronic constraint which requires that at any given time the degrees of belief conform to the probability axioms, and a diachronic constraint which requires that the degree of belief function changes by conditionalization on the evidence that is acquired.[10] In so far as degrees of belief of an agent conform to these strictures, they are deemed to be rational.

By these lights it is rational to give high degrees of belief to miracles; indeed, we have seen that given the priors that many of us — including, presumably, Hume himself — start with, and given various kinds of eyewitness

---

[8] An exception is the hypothetico-deductive case where D entails M.

[9] For a critique of Bayes's attempted justification of his favoured prior probability assignment, see my book (1992).

[10] Some good Bayesians drop the diachronic constraint (e.g. Howson, 2000). But for present purposes it does no harm to impose it.

testimony that it is in principle possible to obtain, it would be irrational not to give a high credibility to miracles. Further, miracles can be used to support rational credence in theological doctrines.

Thus far I have marched shoulder to shoulder with the pro-miracle forces. But now I drop out of the parade, and this for two reasons. First, my degree of belief function—which I immodestly assume to satisfy the Bayesian strictures—disagrees with, for example, Professor Swinburne's (1970, 1979) function—which I have no doubt satisfies the strictures.[11] Second, and more important, I think that these differences are matters of taste in that there is no objective basis to prefer one over the other. One way to find objectivity in the framework of subjective Bayesianism is through an evidence-driven merger of opinion (aka 'washing out of priors'[12]). Such a consensus, however, is hollow unless it is in principle possible that the accumulating evidence produces the merger of opinion by driving the posterior probability to 1 on the true hypotheses and 0 on the false hypotheses. But for theological hypotheses, whose truth values do not supervene on the totality of empirical evidence—no matter how liberally that evidence is construed—the desired convergence to certainty is impossible.

To add some precision to this claim I will introduce a bit of apparatus. Let H and H' be the rival (incompatible) hypotheses at issue. The set of 'possible worlds' allowed by these competing hypotheses are the models $\mathcal{M} := \mathcal{M}^H \cup \mathcal{M}^{H'}$, where $\mathcal{M}^H := \mathrm{mod}(H)$, $\mathcal{M}^{H'} := \mathrm{mod}(H')$, and $\mathrm{mod}(X)$ stands for the set of models that satisfy X. Let $\mathcal{M}_O^H$ and $\mathcal{M}_O^{H'}$ stand for the empirical submodels of $\mathcal{M}^H$ and $\mathcal{M}^{H'}$ respectively. Exactly what is included in these submodels is left vague, but my intent is to be as liberal as possible. For example, they may include not only states of affairs that are directly observable by the unaided senses, but also states of affairs that can only be inferred using elaborate measuring devices and the background theories of these devices. I will say that the truth of the matter as regards H vs H' is *strongly underdetermined* by the empirical iff H and H' have the same empirical content in that there is a one–one map $I : \mathcal{M}^H \to \mathcal{M}^{H'}$ such that for all $m \in \mathcal{M}^H$ the corresponding empirical submodels $m_O \in \mathcal{M}_O^H$ and $I(m)_O \in \mathcal{M}_O^{H'}$ are isomorphic.[13] I call this strong empirical underdetermination because for any empirical situation (submodel) allowed by either hypothesis, the 'same' situation

---

[11] It is only here that I part company with Professor Swinburne, from whom I learned to apply Bayesianism to religious matters.

[12] This phrase is misleading since the likelihoods, which in many cases are just as subjective as the priors, have to wash out too.

[13] A weaker form of empirical underdetermination would obtain if H and H' had some common empirical submodels, i.e. there are $m \in \mathcal{M}^H$ and $m' \in \mathcal{M}^{H'}$ such that m_O and m'_O are isomorphic.

(as expressed in isomorphic submodels) extends to a full model in which H is true and H' is false, and it also extends to a different full model in which H' is true and H is false. For such a pair of rival hypotheses there can be no empirical procedure — Bayesian or non-Bayesian — which is reliable over all the possible worlds $\mathcal{M}$ in detecting the truth of the matter as regards H and H'. For there is no procedure which for each $m \in \mathcal{M}$ operates on the empirical submodel m_O of m and produces the truth values of H and H' in m.[14] If one prefers, the same point can be made in terms of evidence statements. A condition for an evidence statement E to be counted as empirical is that it be treated symmetrically with respect to any pair of isomorphic empirical submodels m_O and m_O': either $m_O \vDash E$ and $m_O' \vDash E$ or else $m_O \vDash {\sim} E$ and $m_O' \vDash {\sim} E$. The set $\mathcal{E}$ of all such evidence statements constitutes what can be called the *empirical evidence matrix*. A possible world m fills in the truth values for the statements in the matrix, the array of which is denoted by $\mathcal{E}^m$. The point now becomes that in cases of strong underdetermination there is no procedure — Bayesian or otherwise — which for each $m \in \mathcal{M}$ looks at some or all of the array $\mathcal{E}^m$ and produces the truth values for H and H' in m.

One might hope that adding background knowledge could make a reliable empirical learning procedure possible since such knowledge will cut down on the set of possible worlds over which the procedure has to be reliable. But this hope is vain when the underdetermination is of the strong variety and when the background knowledge is empirical. Background knowledge may be either propositional or probabilistic. In the former case, the knowledge K reduces the set of possible worlds $\mathcal{M}$ to $\mathcal{M}^* := \mathcal{M}^{*H} \cup \mathcal{M}^{*H'}$ where $\mathcal{M}^{*H} :=$ $\mathrm{mod}(H\&K)$ and $\mathcal{M}^{*H'} := \mathrm{mod}(H'\&K)$. In the case of probabilistic knowledge, $\mathcal{M}^{*H}$ and $\mathcal{M}^{*H'}$ are singled out as subsets of measure one relative to some measure imposed on $\mathcal{M}$. I propose that whatever counts as empirical background knowledge must satisfy the constraint that $I(\mathcal{M}^{*H}) = \mathcal{M}^{*H'}$, where I is the same mapping that exhibits the original strong underdetermination. The idea is that empirical knowledge should be characterizable directly in terms of how it operates on the empirical submodels and, thus, must treat isomorphic empirical submodels symmetrically. Two cases have to be considered. (a) The background knowledge is so strong that either $\mathcal{M}^{*H} = \varnothing$ or $\mathcal{M}^{*H'} = \varnothing$. But in this case the background knowledge amounts to knowledge of the non-empirical since it fixes the truth value of one (or more) of the competing hypotheses to False. (b) $\mathcal{M}^{*H} \neq \varnothing$ and $\mathcal{M}^{*H'} \neq \varnothing$. In this case the constraint

---

[14] More precisely, there are no functions $f_H : \mathcal{M}_O^H \to \{T, F\}$ and $f_{H'} : \mathcal{M}_O^{H'} \to \{T, F\}$ such that for any $m_O \in \mathcal{M}_O^H$ if m_O is the empirical submodel of $m \in \mathcal{M}^H$ then $f(m_O) = T$ (respectively, F) if $m \vDash H$ (respectively, if $m \vDash {\sim} H$), and similarly for $f_{H'}$.

on what counts as empirical background knowledge implies that the strong empirical underdetermination remains intact even after the empirical background knowledge is applied. It follows that only a priori background knowledge of the non-empirical can lead to a reliable empirical procedure for ascertaining the truth of hypotheses subject to strong empirical underdetermination. Whatever its metaphysical legitimacy, such background knowledge is no part of scientific methodology.

To complete my case I need two additional premises: first, that a characteristic feature of religious doctrines is that, even on the most liberal construal of 'the empirical', they are subject to strong empirical underdetermination; and second, that while beliefs or degrees of belief that are formed in a manner that does not reflect a reliable connection between truth and empirical evidence — either because no such reliable connection exists, or else because a reliable connection exists but the belief formation method does not reflect it — may possibly be counted as 'rational', they cannot qualify as objective or scientific. It follows that the project of natural religions is impossible, at least in so far as that project presupposes that theology can proceed, as does science, in gathering empirical evidence and forming objective opinions on the basis of that evidence.

This sour conclusion can be avoided by attacking my additional premises; in particular, it could be claimed either that theological hypotheses are not underdetermined by empirical evidence, or else that I have imposed an impossibly high standard of objectivity on science since high-level hypotheses in theoretical physics are as underdetermined as theological hypotheses. I will not try to respond to the first attack since doing so would involve a substantial discussion of theology, which is something I cannot provide here. As for the second attack, I would turn it around and maintain that, in so far as the hypotheses of theoretical physics are subject to strong empirical underdetermination, theoretical physics is a kind of theology whose high priests should be accorded no more respect, *qua* scientists, than the high priests of Christianity or Buddhism. But I would add that, despite the fact that it is asserted in the literature on scientific realism that underdetermination is a commonplace, I have seen very few interesting examples of it in the history of science. Perhaps the sparsity of examples is due either to a lack of imagination or, more interestingly, to the tacit imposition of a set of criteria for selecting hypotheses to be seriously entertained. In the latter case the proponents of natural religion may be able to argue that what is sauce for the goose is sauce for the gander and that by using analogous selection criteria theologians can match theoretical physicists in overcoming or greatly reducing the scope of underdetermination. These issues merit a vigorous and thorough examination.

9. Conclusion

I trust that I have displeased all parties. I hope to have upset the devotees of Hume's miracles essays by showing that a Bayesian examination reveals Hume's seemingly powerful argument to be a shambles from which little emerges intact, save for posturing and pompous solemnities. At the same time I hope I have given no comfort to the pro-miracle forces. I personally do not give much credibility to religious miracles and religious doctrines. And while I acknowledge that those who do can be just as rational as I am, I suspect that degrees of belief in religious doctrines cannot have an objective status if a necessary condition for such a status is the existence of a reliable procedure for learning, from all possible empirical evidence, the truth values of these doctrines. The proof, or disproof, of this suspicion would, I think, constitute an important contribution to the philosophy of religion.

References

Annet, P. (1744a), *The Resurrection of Jesus Considered, In Answer to the Tryal of the Witnesses. By a Moral Philosopher*, 3rd edn, London: Printed for M. Cooper.

Annet, P. (1744b), *The Resurrection Reconsidered. Being an Answer to Clearer and Others*, London: Printed for the author by M. Cooper.

Anonymous (George Hooper?) (1699), 'A Calculation of the Credibility of Human Testimony', *Philosophical Transactions of the Royal Society* **21**: 359–65.

Babbage, C. (1838), *The Ninth Bridgewater Treatise*. Page references to the 2nd edn, London: Frank Cass and Co., 1967.

Barnard, G. A. (1958), 'Thomas Bayes — A Biographical Note', *Biometrika* **45**: 293–5.

Bayes, T. (1763), 'An Essay Towards Solving a Problem in the Doctrine of Chances', *Philosophical Transactions of the Royal Society* **53**: 370–418. Reprinted in *Biometrika* **45** (1958): 296–315, and in this volume as Appendix, pp. 122–49.

Broad, C. D. (1916–17), 'Hume's Theory of the Credibility of Miracles', *Proceedings of the Aristotelian Society* **17**: 77–94.

Burns, R. M. (1981), *The Great Debate on Miracles, From Joseph Glanville to David Hume*, East Brunswick, NJ: Associated University Presses.

Chandler, S. (1744), *Witnesses of the Resurrection of Jesus Christ Reexamined: And Their Testimony Proved Entirely Consistent*, London: Printed for J. Noon and R. Hett.

Clarke, S. (1705), 'A Discourse Concerning the Unalterable Obligations of Natural Religion, and the Truth and Certainty of the Christian Revelation'. Page references to *The Works of Samuel Clarke*, Vol. 2, pp. 580–733, New York: Garland Publishing, 1978.

Earman, J. (1992), *Bayes or Bust: A Critical Examination of Bayesian Confirmation Theory*, Cambridge, MA: MIT Press.

Earman, J. (2000), *Hume's Abject Failure: The Argument Against Miracles*, New York: Oxford University Press.

Gillies, D. (1991), 'A Bayesian Proof of a Humean Principle', *British Journal for the Philosophy of Science* **42**: 255–6.

Greig, J. Y. T. (ed.) (1932), *Letters of David Hume*, 2 vols, Oxford: Oxford University Press.

Howson, C. (2000), *Hume's Problem: Induction and the Justification of Belief*, Cambridge: Cambridge University Press.

Hume, D. (1748), *Enquiries Concerning Human Understanding And Concerning the Principles of Morals*. Reprinted from the posthumous edition of 1777. Page references to 3rd edn with text revised and notes by P. H. Niddich. Oxford: Clarendon Press.

Jackson, J. (1744), *An Address to Deists*, London: Printed for J. and P. Knapton.

Klibansky, R. and Mossner, E. C. (eds) (1954), *New Letters of David Hume*, Oxford: Clarendon Press.

Laplace, P. S. (1812), *Théorie Analytique des Probabilités*. Page references to the 3rd edn (1820) reprinted in *Oeuvres complètes de Laplace*, Vol. 7, Paris: Gauthier-Villars, 1886.

Laplace, P. S. (1814), *Essai philosophique sur les probabilités*. Page references to F. W. Truscott and F. L. Emory (trans.), *A Philosophical Essay on Probabilities*, New York: Dover, 1951.

Locke, J. (1690), *An Essay Concerning Human Understanding*. Reprinted as *An Essay Concerning Human Understanding, by John Locke*, A. C. Fraser (ed.), New York: Dover, 1959.

Mill, J. S. (1843), *A System of Logic*. Page references to the 8th edn, New York: Harper and Bros, 1874.

Price, R. (1767), *Four Dissertations*. Page references to the 2nd edn 1768, London: A. Millar and T. Cadell.

Sherlock, T. (1728), *Tryal of the Witnesses of the Resurrection of Jesus*. London: J. Roberts. Page references to the 11th edn, London: J. and H. Pembert, 1743.

Sobel, J. H. (1991), 'Hume's Theorem on Testimony Sufficient to Establish a Miracle', *Philosophical Quarterly* **41**: 229–37.

Swinburne, R. (1970), *The Concept of Miracle*, New York: St Martin's Press.

Swinburne, R. (1979), *The Existence of God*, Oxford: Clarendon Press.

Thomas, R. (1924), *Richard Price: Philosopher and Apostle of Liberty*, London: Humphrey Milford.

Tillotson, J. (1664), 'The Wisdom of Being Religious'. Reprinted as Sermon 1 in *The Works of the Most Reverend Dr. John Tillotson*, London: J. Darby, 1728.

West, G. (1747), *Observations on the History and Evidence of the Resurrection of Jesus Christ*, London: Printed for R. Dodsley.

Wilkins, J. (1699), *Of the Principles and Duties of Natural Religion*. Page references to the 5th edn, London: Printed for R. Chiswell, 1704.

Woolston, T. (1727–29), *Six Discourses on the Miracles of Our Savior*, London. Reprinted by Garland Publishing Co., New York, 1979.

6

Propensities May Satisfy Bayes's Theorem

DAVID MILLER

IN THIS PAPER I shall reconsider the question whether the propensity interpretation of probability is, as its name suggests, a genuine interpretation of
the calculus of probability. The main point of interest, but not the only one,
lies in the difficulty of understanding the term $P(A|C)$ when the occurrence A
is temporally or causally anterior to the occurrence C.

In the form in which we are concerned with it here, the propensity interpretation was first proposed by Karl Popper in his paper 'The Propensity
Interpretation of the Calculus of Probability, and the Quantum Theory'
(1957). In the ensuing decades a number of writers took up the idea, developed it, criticized it, and reformulated it in different ways. The reader may
consult Settle (1975) and Gillies (2000), chapter 6, for discriminatory sur-
veys. To my mind the most interesting criticism to date is that pressed by
Humphreys (1985) and Milne (1986), which argues that measures of propensity do not obey Bayes's theorem and that therefore there exists no propensity
interpretation of probability. To this conundrum my *Critical Rationalism*
(1994), chapter 9.5, offered a sketchy resolution. Its main idea was later
thoroughly investigated by McCurdy (1996), a paper that I am largely in
agreement with (though it is suggested incorrectly on p. 106 that I regard
propensities as fundamentally propensities to generate frequencies). But my
curiosity about the propensity interpretation will not let me leave the matter
there.

1. According to the frequency interpretation of probability, probabilities
may under favourable circumstances be attributed to events — that is, to
classes of actual or possible occurrences in the sense of Popper (1934/1959),
§23. Probabilities are completely objective features of the world. According
to several non-physical interpretations, often nicknamed subjectivist, probabilities are attributed to individual statements or propositions or beliefs, and
quantify psychological (or sometimes even logical) aspects of these items.
To most subjectivists it makes good sense to talk about the probability of a

Proceedings of the British Academy, **113**, 111–116. © D. W. Miller, 2002.

statement reporting (truly or falsely) a single dated occurrence of an event, but to most frequentists talk of the probability of a single occurrence is taboo. Satisfied that quantum theory needs to admit singular probabilities, and convinced that a subjectivist construe of these probabilities is incorrect, Popper ventured an alternative objectivist interpretation, in which probability may be ascribed to a single occurrence, actual or possible, of an event, but should not be assumed to measure the frequency with which occurrences of that event will be found to recur. The probability is supposed rather to measure the propensity at a given time for that actual or possible occurrence to come to pass. This interpretation of the probability of an occurrence, actual or possible, is interesting only if the world is not deterministic, for otherwise every propensity has the value 0 or has the value 1.

It is a factual hypothesis that the world is faced at any time with a range of possible ways forward, and that these possibilities may be differently weighted. It is a second hypothesis that these weights or propensities conform to the axioms of the calculus of probability. But if propensities really are probabilities, and are sufficiently well behaved, then a bridge between propensities and frequencies is provided by the laws of large numbers. Statements of propensity are in principle testable.

2.   The difficulty to be discussed here concerns not the interpretation of the term $P_t(A)$, the (absolute) propensity at time t for the actual or possible occurrence A to be realized, but the term $P_t(A|C)$, where A and C are both occurrences. Recall that an occurrence is here taken to be what is described by a basic statement, a singular statement equipped with coordinates of time and place. Intuitively we may read the term $P_t(A|C)$ as the propensity at time t for the occurrence A to be realized given that the occurrence C is realized. If C precedes A in time, this presents no extraordinary difficulty. But if C follows A, or is simultaneous (or even identical) with A, then it appears that there is no propensity for A to be realized given that C is realized; for either A has been realized already, or it has not been realized and never will be. In a late work Popper wrote that 'a propensity zero means *no* propensity' (1990, p. 13). If the converse holds too, we may conclude that $P_t(A|C)$ has the value zero unless C precedes A in time. But it is easy to construct examples in which none of $P_t(C|A)$, $P_t(A)$, and $P_t(C)$ is zero. The simplest version of Bayes's Theorem, to the effect that $P_t(A|C) = P_t(C|A)P_t(A)/P_t(C)$, is thereby violated.

3.   The objection involves a subtle misreading into the phrase 'given that' of an inappropriate temporal reference. Suppose that A is an occurrence that is realized, if it is realized at all, at time t', and that C is an occurrence that is realized, if it is realized at all, at time t''. Talk of the propensity at time t for A to be realized (at time t') given that C is realized (at time t'') does not mean that the realization of C at time t'' is supposed to be given at time t'. It means that the realization of C at time t'' is supposed to be given at time t. Of course if t is earlier than t'', then this supposition is subjunctive. But provided that t is earlier than t', there is no difficulty in principle in attributing a positive value to $P_t(A_{t'}|C_{t''})$. Note that if t'' too is earlier than t', and C comes to pass at t'', then there is an innocuous sense in which the occurrence of C is given at t' — by the time t' is reached, C has been realized. This is not the sense of the phrase 'given that' that is central to the theory of relative probability.

Only if t is earlier than t'' can $P_t(A_{t'}|C_{t''})$ differ from $P_t(A_{t'})$, the absolute propensity at t for A to be realized at t'. We may set aside as uninteresting the case in which t is not earlier than t''. Now it should be obvious that to suppose at t that C comes to pass at t'' is not to suppose incoherently that every occurrence dated between t and t'' also comes to pass; it is not even to suppose at t that we are already at t''. Provided therefore that t is earlier than t', to suppose at t that C comes to pass at t'' is not to suppose either that A comes to pass at t' or that it does not come to pass at t', even if t' is earlier than t''. In consequence there is, if t' is earlier than t'', nothing in principle that disallows $P_t(A_{t'}|C_{t''})$ from taking any value greater than zero. Of course, the value of $P_t(A_{t'}|C_{t''})$ will be either zero or unity unless t is earlier than t'.

4. Table 1, in which '$x < z$' is shorthand for 'x is earlier than z', summarizes the different situations that may arise.

In rows 1 and 2, A has already come to pass, or failed to come to pass, at time t', which is not later than time t. Row 3 is the tricky case that we have just considered, in which the time of the conditioning occurrence C is not earlier than the time of the conditioned occurrence A. Row 4 is the standard case, in which the time of the conditioning occurrence C is earlier than the time of the conditioned occurrence A. In rows 5 and 6, either C has already come to pass at time t'', which is not later than time t, in which case it is a necessity at time t, and hence $P_t(A_{t'}|C_{t''}) = P_t(A_{t'})$; or it has failed to come to pass at time t'', in which case it is an impossibility at time t, and hence $P_t(A_{t'}|C_{t''}) = 1$. One possibility in row 5, of course, is that A has failed to come to pass at time t', so that $P_t(A_{t'}|C_{t''}) = P_t(A_{t'}) = 0$, as on rows 1 and 2.

5. In chapter 9.5 of my (1994) I expressed the analysis contained in §3 and row 3 above by saying that $P_t(A_{t'}|C_{t''})$ is the propensity of the world at time t to develop into a world in which A comes to pass at time t', given that it (the world at time t) develops into a world in which C comes to pass at time t''

David Miller

Table 1

$t' \le t \le t''$	$P_t(A_{t'}|C_{t''}) = 0$ or $P_t(A_{t'}|C_{t''}) = 1$
$t' \le t'' < t$	$P_t(A_{t'}|C_{t''}) = 0$ or $P_t(A_{t'}|C_{t''}) = 1$
$t \le t' \le t''$	$P_t(A_{t'}|C_{t''})$ may take any value $\in [0,1]$
$t \le t'' < t'$	$P_t(A_{t'}|C_{t''})$ may take any value $\in [0,1]$
$t'' \le t' \le t$	$P_t(A_{t'}|C_{t''}) = P_t(A_{t'})$ or $P_t(A_{t'}|C_{t''}) = 1$
$t'' \le t < t'$	$P_t(A_{t'}|C_{t''}) = P_t(A_{t'})$ or $P_t(A_{t'}|C_{t''}) = 1$

(p. 189). That is perfectly correct. It is sufficiently correct too to say, as I did there, that 'the causal pressure' is from time t to time t', not from time t'' to time t'. Humphreys has pointed out to me in correspondence that some other details of my earlier treatment are not entirely satisfactory. In particular, it was misleading to suggest that the conditional propensity $P_t(A_{t'}|C_{t''})$ can most easily be thought of as what the propensity for A to take place at t' would be at t if something were to happen at t to guarantee that C takes place at t''. This cannot be quite right, because we cannot suppose there to exist any occurrence at t that guarantees C's taking place at t'' and nothing more besides. In this respect the propensity interpretation is less flexible than most other interpretations.

6. Some writers, especially Fetzer (1981), have construed propensities as partial causes, the term $P(A|C)$ being understood to grade the strength of the propensity of the conditions C at time t to produce the event A at some time later than t (here I paraphrase a report in Gillies, 2000, p. 135). The problem raised by Humphreys and Milne is unavoidable in this version of the propensity interpretation, and indeed Fetzer is not concerned to avoid it. He squarely acknowledges that if causes are supposed to reside in sets of conditions, then probabilities make good measures of partial causation or propensity in some cases only. This manoeuvre leaves the calculus of probability only partially interpreted; and it is therefore considerably less appealing if one's main interest is probability than it may be if one's main interest is partial causation (for which Fetzer offers an alternative formalism). For even one of the central axioms of the calculus of probability, axiom A3 of Popper (1959), appendix *v, which states that $P(x|x) = P(z|z)$, is disallowed if the two arguments of the probability functor are required to refer to different times.

In any case, Fetzer's dramatic response can be avoided if we move the locus of the propensity from the conditioning term C to the time t. There is no dispute about the significance of the time t at which the propensity is evaluated, since it is agreed on all sides that propensities can in general change in time.

7. Now it may be that few will wish to credit a time, or even the state of the world at a time, with causal efficacy. The general tenor of most work on the propensity interpretation, from Popper's introductory papers to the discussions in Settle (1974, 1975) and beyond, has been to treat propensities as non-deterministic dispositions, residing either in objects or in local situations or experimental arrangements. (Later, for example in the preface to *Quantum Theory and the Schism in Physics,* 1982, Popper acknowledged the need to relativize propensity to a time.) The motive to localize propensities in experimental arrangements or generating conditions is prompted, to be sure, by the admirable desire to make statements of propensity universal rather than singular, and thereby to make them amenable to testing. This motive is very evident in Gillies (2000), who criticizes the frequency interpretation of von Mises for being too operationalistic (pp. 137f.), and the version of the propensity interpretation defended here for being too metaphysical (pp. 127f.). But on this point Gillies is wrong. It is true that statements of the form $P_t(A_{t'}|C_{t''}) = q$ are not testable in isolation, but in that respect they are exactly like statements of instantaneous velocity or acceleration in classical mechanics. That does not make analyses of classical dynamical systems untestable. In the same way, if we can formulate universal statements about how propensities change, or do not change, we can test statements of propensity as well as we can test any other statements of probability, through the medium of frequencies.

8. In summary, there exists an intelligible interpretation of the calculus of probability as a calculus of propensities. It is not obviously a correct interpretation; as stressed in **1**, para. 2, it is a factual matter whether propensities obey the calculus of probability. What is clear is that this calculus is not a calculus of partial causation. Indeed, I think, we shall eventually have to discard altogether the lazy idea that propensities are generalized dispositions or partial causes, though I know that Popper himself sometimes explicitly made this identification (for example, in *A World of Propensities,* 1990, p. 20). His earlier idea, never abandoned, that the propensities with which the propensity interpretation deals are generalized forces, strikes me as a much better candidate for the truth. But this is not the place to defend the doctrine that forces (which are sometimes observable) are more effective than causes (which are never observable).

References

Fetzer, J. (1981), *Scientific Knowledge. Causation, Explanation, and Corroboration,* Dordrecht: D. Reidel Publishing Company.

Gillies, D. A. (2000), *Philosophical Theories of Probability*, London and New York: Routledge.

Humphreys, P. W. (1985), 'Why Propensities Cannot be Probabilities', *Philosophical Review* **XCIV**: 557–70.

McCurdy, C. S. I. (1996), 'Humphreys's Paradox and the Interpretation of Inverse Conditional Propensities', *Synthese* **108**: 105–25.

Miller, D. W. (1994), *Critical Rationalism. A Restatement & Defence*, Chicago & La Salle: Open Court Publishing Company.

Milne, P. (1986), 'Can There Be a Realist Single-Case Interpretation of Probability?', *Erkenntnis* **25**: 129–32.

Popper, K. R. (1934), *Logik der Forschung*, Vienna: Julius Springer Verlag.

Popper, K. R. (1957), 'The Propensity Interpretation of the Calculus of Probability, and the Quantum Theory', in S. Körner (ed.), *Observation & Interpretation in the Philosophy of Physics*, London: Butterworth & Company Ltd, pp. 65–70. Reprinted as selection 15 of D.W. Miller (ed.), *Popper Selections*, Princeton: Princeton University Press, 1985.

Popper, K. R. (1959), *The Logic of Scientific Discovery*, London: Hutchinson & Company Ltd. Enlarged English translation of Popper (1934).

Popper, K. R. (1982), *Quantum Theory & the Schism in Physics*, London: Hutchinson & Company Ltd.

Popper, K. R. (1990), *A World of Propensities*, Bristol: Thoemmes Antiquarian Books Ltd.

Settle, T. W. (1974), 'Induction and Probability Unfused', in P. A. Schilpp (ed.), *The Philosophy of Karl Popper*, La Salle: Open Court Publishing Company, pp. 697–749.

Settle, T. W. (1975), 'Presuppositions of Propensity Theories of Probability', in G. Maxwell and R. M. Anderson, Jr (eds), *Induction, Probability, & Confirmation*, Minneapolis: University of Minnesota Press, pp. 388–415.

APPENDIX

'An Essay Towards Solving a Problem in the Doctrine of Chances' by Thomas Bayes, presented to the Royal Society by Richard Price*

Thomas Bayes — a biographical note by G. A. Barnard

BAYES'S PAPER, reproduced in the following pages, must rank as one of the most famous memoirs in the history of science and the problem it discusses is still the subject of keen controversy. The intellectual stature of Bayes himself is measured by the fact that it is still of scientific as well as historical interest to know what Bayes had to say on the questions he raised. And yet such are the vagaries of historical records, that almost nothing is known about the personal history of the man. *The Dictionary of National Biography*, compiled at the end of the last century, when the whole theory of probability was in temporary eclipse in England, has an entry devoted to Bayes's father, Joshua Bayes, F.R.S., one of the first six Nonconformist ministers to be publicly ordained as such in England, but it has nothing on his much more distinguished son. Indeed, the note on Thomas Bayes which is to appear in the forthcoming new edition of the *Encyclopedia Britannica* will apparently be the first biographical note on Bayes to appear in a work of general reference since the *Imperial Dictionary of Universal Biography* was published in Glasgow in 1865. And in treatises on the history of mathematics, such as that of Loria (1933) and Cantor (1908), notice is taken of his contributions to probability theory and to mathematical analysis, but biographical details are lacking.

The Reverend Thomas Bayes, F.R.S., author of the first expression in precise, quantitative form of one of the modes of inductive inference, was

---

* This version of Bayes's paper, together with G. A. Barnard's historical introduction, is reprinted with permission from *Biometrika* **45** (1958): 293–315.

born in 1702, the eldest son of Ann Bayes and Joshua Bayes, F.R.S. He was educated privately, as was usual with Nonconformists at that time, and from the fact that when Thomas was 12 Bernoulli wrote to Leibniz that 'poor de Moivre' was having to earn a living in London by teaching mathematics, we are tempted to speculate that Bayes may have learned mathematics from one of the founders of the theory of probability. Eventually Thomas was ordained, and began his ministry by helping his father, who was, at the time stated, minister of the Presbyterian meeting house in Leather Lane, off Holborn. Later the son went to minister in Tunbridge Wells at the Presbyterian Chapel on Little Mount Sion which had been opened on 1 August 1720. It is not known when Bayes went to Tunbridge Wells, but he was not the first to minister on Little Mount Sion, and he was certainly there in 1731, when he produced a tract entitled 'Divine Benevolence, or an attempt to prove that the Principal End of the Divine Providence and Government is the happiness of His Creatures'. The tract was published by John Noon and copies are in Dr Williams's library and the British Museum. The following is a quotation:

> [p. 22]: I don't find (I am sorry to say it) any necessary connection between mere intelligence, though ever so great, and the love or approbation of kind and beneficent actions.

Bayes argued that the principal end of the Deity was the happiness of His creatures, in opposition to Balguy and Grove who had, respectively, maintained that the first spring of action of the Deity was Rectitude, and Wisdom.

In 1736 John Noon published a tract entitled 'An Introduction to the Doctrine of Fluxions, and a Defence of the Mathematicians against the objections of the Author of the Analyst'. De Morgan (1860) says: 'This very acute tract is anonymous, but it was always attributed to Bayes by the contemporaries who write in the names of the authors as I have seen in various copies, and it bears his name in other places.' The ascription to Bayes is accepted also in the British Museum catalogue.

From the copy in Dr Williams's library we quote:

> [p. 9]: It is not the business of the Mathematician to dispute whether quantities do in fact ever vary in the manner that is supposed, but only whether the notion of their doing so be intelligible; which being allowed, he has a right to take it for granted, and then see what deductions he can make from that supposition. It is not the business of a Mathematician to show that a strait line or circle can be drawn, but he tells you what he means by these; and if you understand him, you may proceed further with him; and it would not be to the purpose to object that there is no such thing in nature as a true strait line or perfect circle, for this is none of his

concern: he is not inquiring how things are in matter of fact, but supposing things to be in a certain way, what are the consequences to be deduced from them; and all that is to be demanded of him is, that his suppositions be intelligible, and his inferences just from the suppositions he makes.

[p. 48]: He [i.e. the Analyst = Bishop Berkeley] represents the disputes and controversies among mathematicians as disparaging the evidence of their methods: and, Query 51, he represents Logics and Metaphysics as proper to open their eyes, and extricate them from their difficulties. Now were ever two things thus put together? If the disputes of the professors of any science disparage the science itself, Logics and Metaphysics are much more to be disparaged than Mathematics; why, therefore, if I am half blind, must I take for my guide one that can't see at all?

[p. 50]: So far as Mathematics do not tend to make men more sober and rational thinkers, wiser and better men, they are only to be considered as an amusement, which ought not to take us off from serious business.

This tract may have had something to do with Bayes's election, in 1742, to Fellowship of the Royal Society, for which his sponsors were Earl Stanhope, Martin Folkes, James Burrow, Cromwell Mortimer, and John Eames.

William Whiston, Newton's successor in the Lucasian Chair at Cambridge, who was expelled from the University for Arianism, notes in his Memoirs (p. 390) that 'on August the 24th this year 1746, being Lord's Day, and St Bartholomew's Day, I breakfasted at Mr Bay's, a dissenting Minister at Tunbridge Wells, and a Successor, though not immediate, to Mr Humphrey Ditton, and like him a very good mathematician also'. Whiston goes on to relate what he said to Bayes, but he gives no indication that Bayes made reply.

According to Strange (1949) Bayes wished to retire from his ministry as early as 1749, when he allowed a group of Independents to bring ministers from London to take services in his chapel week by week, except for Easter, 1750, when he refused his pulpit to one of these preachers; and in 1752 he was succeeded in his ministry by the Rev. William Johnston, A.M., who inherited Bayes's valuable library. Bayes continued to live in Tunbridge Wells until his death on 17 April 1761. His body was taken to be buried, with that of his father, mother, brothers and sisters, in the Bayes and Cotton family vault in Bunhill Fields, the Nonconformist burial ground by Moorgate. This cemetery also contains the grave of Bayes's friend, the Unitarian Rev. Richard Price, author of the *Northampton Life Table* and object of Burke's oratory and invective in *Reflections on the French Revolution*, and the graves of John Bunyan, Samuel Watts, Daniel Defoe, and many other famous men.

Bayes's will, executed on 12 December 1760, shows him to have been a man of substance. The bulk of his estate was divided among his brothers, sisters, nephews and cousins, but he left £200 equally between 'John Boyl late preacher at Newington and now at Norwich, and Richard Price now

I suppose preacher at Newington Green'. He also left 'To Sarah Jeffery daughter of John Jeffery, living with her father at the corner of Fountains Lane near Tonbridge Wells, £500, and my watch made by Elliott and all my linen and wearing apparell and household stuff.'

Apart from the tracts already noted, and the celebrated Essay reproduced here, Bayes wrote a letter on Asymptotic Series to John Canton, published in the *Philosophical Transactions of the Royal Society* (1763, pp. 269–271). His mathematical work, though small in quantity, is of the very highest quality; both his tract on fluxions and his paper on asymptotic series contain thoughts which did not receive as clear expression again until almost a century had elapsed.

Since copies of the volume in which Bayes's essay first appeared are not rare, and copies of a photographic reprint issued by the Department of Agriculture, Washington, D.C., U.S.A., are fairly widely dispersed, the view has been taken that in preparing Bayes's paper for publication here some editing is permissible. In particular, the notation has been modernized, some of the archaisms have been removed and what seem to be obvious printer's errors have been corrected. Sometimes, when a word has been omitted in the original, a suggestion has been supplied, enclosed in square brackets. Otherwise, however, nothing has been changed, and we hope that while the present text should in no sense be regarded as definitive, it will be easier to read on that account. All the work of preparing the text for the printer was most painstakingly and expertly carried out by Mr M. Gilbert, B.Sc., A.R.C.S. Thanks are also due to the Royal Society for permission to reproduce the Essay in its present form.

In writing the biographical notes the present author has had the friendly help of many persons, including especially Dr A. Fletcher and Mr R. L. Plackett, of the University of Liverpool, Mr J. F. C. Willder, of the Department of Pathology, Guy's Hospital Medical School, and Mr M. E. Ogborn, F.I.A., of the Equitable Life Assurance Society. He would also like to thank Sir Ronald Fisher, for some initial prodding which set him moving, and Prof. E. S. Pearson, for patient encouragement to see the matter through to completion.

References

Anderson, J. G. (1941). *Mathematical Gazette*, **25**, 160–2.
Cantor, M. (1908). *Geschichte der Mathematik*, vol. IV. (Article by Netto.)
De Morgan, A. (1860). *Notes and Queries*, 7 Jan. 1860.

Loria, G. (1933). *Storia delle Matematiche*, vol. III. Turin.

Mackenzie, M. (Ed.) (1865). *Imperial Dictionary of Universal Biography*, 3 vols. Glasgow.

Strange, C. H. (1949). *Nonconformity in Tunbridge Wells*. Tunbridge Wells.

The Gentleman's Magazine (1761). **31**, 188.

Notes and Queries (1941). 19 April.

[Since this biographical note was written, Mr O. B. Sheynin has suggested that reference should be made to a second contribution from Price, "Supplement to the Essay on a Problem in the Doctrine of Chances" (*Phil. Trans.* 1765, **54**, 296–335). This is concerned with improving approximations made in the main Essay. Ed.]

AN ESSAY TOWARDS SOLVING A PROBLEM IN THE DOCTRINE OF CHANCES

By the late Rev. Mr BAYES, F.R.S.

Communicated by Mr Price, in a Letter to John Canton, A.M., F.R.S.

Read 23 December 1763

Dear Sir,

I now send you an essay which I have found among the papers of our deceased friend Mr Bayes, and which, in my opinion, has great merit, and well deserves to be preserved. Experimental philosophy, you will find, is nearly interested in the subject of it; and on this account there seems to be particular reason for thinking that a communication of it to the Royal Society cannot be improper.

He had, you know, the honour of being a member of that illustrious Society, and was much esteemed by many in it as a very able mathematician. In an introduction which he has writ to this Essay, he says, that his design at first in thinking on the subject of it was, to find out a method by which we might judge concerning the probability that an event has to happen, in given circumstances, upon supposition that we know nothing concerning it but that, under the same circumstances, it has happened a certain number of times, and failed a certain other number of times. He adds, that he soon perceived that it would not be very difficult to do this, provided some rule could be found according to which we ought to estimate the chance that the probability for the happening of an event perfectly unknown, should lie between any two named degrees of probability, antecedently to any experiments made about it; and that it appeared to him that the rule must be to suppose the chance the same that it should lie between any two equidifferent degrees; which, if it were allowed, all the rest might be easily calculated in the common method of proceeding in the doctrine of chances. Accordingly, I find among his papers a very ingenious solution of this problem in this way. But he afterwards considered, that the *postulate* on which he had argued might not perhaps be looked upon by all as reasonable; and therefore he chose to lay down in another form the proposition in which he thought the solution of the problem is contained, and in a *scholium* to subjoin the reasons why he thought so, rather than to take into his mathematical reasoning any thing that might admit dispute. This, you will observe, is the method which he has pursued in this essay.

Every judicious person will be sensible that the problem now mentioned is by no means merely a curious speculation in the doctrine of chances, but necessary to be solved in order to [provide] a sure foundation for all our reasonings concerning past facts, and what is likely to be hereafter. Common sense is indeed sufficient to shew us that, from the observation of what has in former instances been the consequence of a certain cause or action, one may make a judgment what is likely to be the consequence of it another time, and that the larger [the] number of experiments we have to support a conclusion, so much the more reason we have to take it for granted. But it is certain that we cannot determine, at least not to any nicety, in what degree repeated experiments confirm a conclusion, without the particular discussion of the beforementioned problem; which, therefore, is necessary to be considered by any one who would give a clear account of the strength of *analogical* or *inductive reasoning*; concerning which, at present, we seem to know little more than that it does sometimes in fact convince us, and at other times not; and that, as it is the means of [a]cquainting us with many truths, of which otherwise we must have been ignorant; so it is, in all probability, the source of many errors, which perhaps might in some measure be avoided, if the force that this sort of reasoning ought to have with us were more distinctly and clearly understood.

These observations prove that the problem enquired after in this essay is no less important than it is curious. It may be safely added, I fancy, that it is also a problem that has never before been solved. Mr De Moivre, indeed, the great improver of this part of mathematics, has in his *Laws of Chance*,* after Bernoulli, and to a greater degree of exactness, given rules to find the probability there is, that if a very great number of trials be made concerning any event, the proportion of the number of times it will happen, to the number of times it will fail in those trials, should differ less than by small assigned limits from the proportion of the probability of its happening to the probability of its failing in one single trial. But I know of no person who has shewn how to deduce the solution of the converse problem to this; namely, 'the number of times an unknown event has happened and failed being given, to find the chance that the probability of its happening should lie somewhere between any two named degrees of probability.' What Mr De Moivre has done therefore cannot be thought sufficient to make the consideration of this point unnecessary; especially, as the rules he has given are not pretended to be rigorously exact, except on supposition that the number of trials made are infinite; from

---

* See Mr De Moivre's *Doctrine of Chances*, p. 243, etc. He has omitted the demonstrations of his rules, but these have been since supplied by Mr Simpson at the conclusion of his treatise on *The Nature and Laws of Chance*.

whence it is not obvious how large the number of trials must be in order to make them exact enough to be depended on in practice.

Mr De Moivre calls the problem he has thus solved, the hardest that can be proposed on the subject of chance. His solution he has applied to a very important purpose, and thereby shewn that those are much mistaken who have insinuated that the Doctrine of Chances in mathematics is of trivial consequence, and cannot have a place in any serious enquiry.* The purpose I mean is, to shew what reason we have for believing that there are in the constitution of things fixt laws according to which events happen, and that, therefore, the frame of the world must be the effect of the wisdom and power of an intelligent cause; and thus to confirm the argument taken from final causes for the existence of the Deity. It will be easy to see that the converse problem solved in this essay is more directly applicable to this purpose: for it shews us, with distinctness and precision, in every case of any particular order or recurrency of events, what reason there is to think that such recurrency or order is derived from stable causes or regulations in nature, and not from any of the irregularities of chance.

The two last rules in this essay are given without the deductions of them. I have chosen to do this because these deductions, taking up a good deal of room, would swell the essay too much; and also because these rules, though of considerable use, do not answer the purpose for which they are given as perfectly as could be wished. They are however ready to be produced, if a communication of them should be thought proper. I have in some places writ short notes, and to the whole I have added an application of the rules in the essay to some particular cases, in order to convey a clearer idea of the nature of the problem, and to shew how far the solution of it has been carried.

I am sensible that your time is so much taken up that I cannot reasonably expect that you should minutely examine every part of what I now send you. Some of the calculations, particularly in the Appendix, no one can make without a good deal of labour. I have taken so much care about them, that I believe there can be no material error in any of them; but should there be any such errors, I am the only person who ought to be considered as answerable for them.

Mr Bayes has thought fit to begin his work with a brief demonstration of the general laws of chance. His reason for doing this, as he says in his introduction, was not merely that his reader might not have the trouble of searching elsewhere for the principles on which he has argued, but because he did not know whither to refer him for a clear demonstration of them. He has also made an apology for the peculiar definition he has given of the word *chance* or *probability*. His design herein was to cut off all dispute about the meaning

---

* See his *Doctrine of Chances*, p. 252, etc.

of the word, which in common language is used in different senses by persons of different opinions, and according as it is applied to *past* or *future* facts. But whatever different senses it may have, all (he observes) will allow that an expectation depending on the truth of any *past* fact, or the happening of any *future* event, ought to be estimated so much the more valuable as the fact is more likely to be true, or the event more likely to happen. Instead therefore, of the proper sense of the word *probability*, he has given that which all will allow to be its proper measure in every where the word is used. But it is time to conclude this letter. Experimental philosophy is indebted to you for several discoveries and improvements; and, therefore, I cannot help thinking that there is a peculiar propriety in directing to you the following essay and appendix. That your enquiries may be rewarded with many further successes, and that you may enjoy every valuable blessing, is the sincere wish of, Sir,

your very humble servant,

Newington-Green,   Richard Price
10 *November* 1763

PROBLEM

Given the number of times in which an unknown event has happened and failed: *Required* the chance that the probability of its happening in a single trial lies somewhere between any two degrees of probability that can be named.

Section I

DEFINITION 1. Several events are *inconsistent*, when if one of them happens, none of the rest can.

2. Two events are *contrary* when one, or other of them must; and both together cannot happen.

3. An event is said to *fail*, when it cannot happen; or, Which comes to the same thing, when its contrary has happened.

4. An event is said to be determined when it has either happened or failed.

5. The *probability of any event* is the ratio between the value at which an expectation depending on the happening of the event ought to be computed, and the value of the thing expected upon its happening.

6. By *chance* I mean the same as probability.

7. Events are independent when the happening of any one of them does neither increase nor abate the probability of the rest.

Prop. 1

When several events are inconsistent the probability of the happening of one or other of them is the sum of the probabilities of each of them.

Suppose there be three such events, and whichever of them happens I am to receive N, and that the probability of the 1st, 2nd, and 3rd are respectively a/N, b/N, c/N. Then (by the definition of probability) the value of my expectation from the 1st will be a, from the 2nd b, and from the 3rd c. Wherefore the value of my expectations from all three will be $a + b + c$. But the sum of my expectations from all three is in this case an expectation of receiving N upon the happening of one or other of them. Wherefore (by definition 5) the probability of one or other of them is $(a + b + c)/N$ or $a/N + b/N + c/N$. The sum of the probabilities of each of them.

COROLLARY. If it be certain that one or other of the three events must happen, then $a + b + c = N$. For in this case all the expectations together amounting to a certain expectation of receiving N, their values together must be equal to N. And from hence it is plain that the probability of an event added to the probability of its failure (or of its contrary) is the ratio of equality. For these are two inconsistent events, one of which necessarily happens. Wherefore if the probability of an event is P/N that of its failure will be $(N - P)/N$.

Prop. 2

If a person has an expectation depending on the happening of an event, the probability of the event is to the probability of its failure as his loss if it fails to his gain if it happens.

Suppose a person has an expectation of receiving N, depending on an event the probability of which is P/N. Then (by definition 5) the value of his expectation is P, and therefore if the event fail, he loses that which in value is P; and if it happens he receives N, but his expectation ceases. His gain therefore is $N - P$. Likewise since the probability of the event is P/N, that of its failure (by corollary prop. 1) is $(N - P)/N$. But P/N is to $(N - P)/N$ as P is to $N - P$, i.e. the probability of the event is to the probability of its failure, as his loss if it fails to his gain if it happens.

Prop. 3

The probability that two subsequent events will both happen is a ratio compounded of the probability of the 1st, and the probability of the 2nd on supposition the 1st happens.

Suppose that, if both events happen, I am to receive N, that the probability both will happen is P/N, that the 1st will is a/N (and consequently that the

1st will not is $(N - a)/N$ and that the 2nd will happen upon supposition the 1st does is b/N. Then (by definition 5) P will be the value of my expectation, which will become b if the 1st happens. Consequently if the 1st happens, my gain by it is $b - P$, and if it fails my loss is P. Wherefore, by the foregoing proposition, a/N is to $(N - a)/N$, i.e. a is to $N - a$ as P is to $b - P$. Wherefore (*componendo inverse*) a is to N as P is to b. But the ratio of P to N is compounded of the ratio of P to b, and that of b to N. Wherefore the same ratio of P to N is compounded of the ratio of a to N and that of b to N, i.e. the probability that the two subsequent events will both happen is compounded of the probability of the 1st and the probability of the 2nd on supposition the 1st happens.

COROLLARY. Hence if of two subsequent events the probability of the 1st be a/N, and the probability of both together be P/N, then the probability of the 2nd on supposition the 1st happens is P/a.

Prop. 4

If there be two subsequent events to be determined every day, and each day the probability of the 2nd is b/N and the probability of both P/N, and I am to receive N if both the events happen the first day on which the 2nd does; I say, according to these conditions, the probability of my obtaining N is P/b. For if not, let the probability of my obtaining N be x/N and let y be to x as $N - b$ to N. Then since x/N is the probability of my obtaining N (by definition 1) x is the value of my expectation. And again, because according to the foregoing conditions the first day I have an expectation of obtaining N depending on the happening of both the events together, the probability of which is P/N, the value of this expectation is P. Likewise, if this coincident should not happen I have an expectation of being reinstated in my former circumstances, i.e. of receiving that which in value is x depending on the failure of the 2nd event the probability of which (by cor. prop. 1) is $(N - b)/N$ or y/x, because y is to x as $N - b$ to N. Wherefore since x is the thing expected and y/x the probability of obtaining it, the value of this expectation is y. But these two last expectations together are evidently the same with my original expectation, the value of which is x, and therefore $P + y = x$. But y is to x as $N - b$ is to N. Wherefore x is to P as N is to b, and x/N (the probability of my obtaining N) is P/b.

COR. Suppose after the expectation given me in the foregoing proposition, and before it is at all known whether the 1st event has happened or not, I should find that the 2nd event has happened; from hence I can only infer that the event is determined on which my expectation depended, and have no reason to esteem the value of my expectation either greater or less than it was before. For if I have reason to think it less, it would be reasonable for me to give something to be reinstated in my former circumstances, and this over and over again as often as I should be informed that the 2nd event had happened, which is evidently absurd. And the like absurdity plainly follows if you say I ought to set a greater value on my expectation than before, for then it would be reasonable for me to refuse something if offered me upon condition I would relinquish it, and be reinstated in my former circumstances; and this likewise over and over again as often as (nothing being known concerning the 1st event) it should appear that the 2nd had happened. Notwithstanding therefore this discovery that the 2nd event has happened, my expectation ought to be esteemed the same in value as before, i.e. x, and consequently the probability of my obtaining N is (by definition 5) still x/N or P/b.* But after this discovery the probability of my obtaining N is the probability that the 1st of two subsequent events has happened upon the supposition that the 2nd has, whose probabilities were as before specified. But the probability that an event has happened is the same as the probability I have to guess right if I guess it has happened. Wherefore the following proposition is evident.

Prop. 5

If there be two subsequent events, the probability of the 2nd b/N and the probability of both together P/N, and it being first discovered that the 2nd event has happened, from hence I guess that the 1st event has also happened, the probability I am in the right is P/b.†

---

* What is here said may perhaps be a little illustrated by considering that all that can be lost by the happening of the 2nd event is the chance I should have had of being reinstated in my former circumstances, if the event on which my expectation depended had been determined in the manner expressed in the proposition. But this chance is always as much *against* me as it is *for* me. If the 1st event happens, it is *against* me, and equal to the chance for the 2nd event's failing. If the 1st event does not happen, it is *for* me, and equal also to the chance for the 2nd event's failing. The loss of it, therefore, can be no disadvantage.

† What is proved by Mr Bayes in this and the preceding proposition is the same with the answer to the following question. What is the probability that a certain event, when it happens, will be accompanied with another to be determined at the same time? In this case, as one of the events is given, nothing can be due for the expectation of it: and, consequently, the value of an expectation depending on the happening of both events must be the same with the value of an expectation depending on the happening of one of them. In other words; the probability that, when one of two events happens, the other will, is the same with the probability of this other. Call x then the probability of this other, and if b/N be the probability of the given event, and p/N the probability of both, because $p/N = (b/N) \times x$, $x = p/b =$ the probability mentioned in these propositions.

Prop. 6

The probability that several independent events shall all happen is a ratio compounded of the probabilities of each.

For from the nature of independent events, the probability that any one happens is not altered by the happening or failing of any of the rest, and consequently the probability that the 2nd event happens on supposition the 1st does is the same with its original probability; but the probability that any two events happen is a ratio compounded of the probability of the 1st event, and the probability of the 2nd on supposition the 1st happens by prop. 3. Wherefore the probability that any two independent events both happen is a ratio compounded of the probability of the 1st and the probability of the 2nd. And in like manner considering the 1st and 2nd events together as one event; the probability that three independent events all happen is a ratio compounded of the probability that the two 1st both happen and the probability of the 3rd. And thus you may proceed if there be ever so many such events; from whence the proposition is manifest.

Cor. 1. If there be several independent events, the probability that the 1st happens, the 2nd fails, the 3rd fails and the 4th happens, etc. is a ratio compounded of the probability of the 1st, and the probability of the failure of the 2nd, and the probability of the failure of the 3rd, and the probability of the 4th, etc. For the failure of an event may always be considered as the happening of its contrary.

Cor. 2. If there be several independent events, and the probability of each one be a, and that of its failing be b, the probability that the 1st happens and the 2nd fails, and the 3rd fails and the 4th happens, etc. will be $abba$, etc. For, according to the algebraic way of notation, if a denote any ratio and b another, $abba$ denotes the ratio compounded of the ratios a, b, b, a. This corollary therefore is only a particular case of the foregoing.

Definition. If in consequence of certain data there arises a probability that a certain event should happen, its happening or failing, in consequence of these data, I call its happening or failing in the 1st trial. And if the same data be again repeated, the happening or failing of the event in consequence of them I call its happening or failing in the 2nd trial; and so on as often as the same data are repeated. And hence it is manifest that the happening or failing of the same event in so many diffe[rent] trials, is in reality the happening or failing of so many distinct independent events exactly similar to each other.

Prop. 7

If the probability of an event be a, and that of its failure be b in each single trial, the probability of its happening p times, and failing q times in

$p + q$ trials is $Ea^p b^q$ if E be the coefficient of the term in which occurs $a^p b^q$ when the binomial $(a + b)^{p+q}$ is expanded.

For the happening or failing of an event in different trials are so many independent events. Wherefore (by cor. 2 prop. 6) the probability that the event happens the 1st trial, fails the 2nd and 3rd, and happens the 4th, fails the 5th, etc. (thus happening and failing till the number of times it happens be p and the number it fails be q) is *abbab* etc. till the number of *a*'s be p and the number of *b*'s be q, that is; 'tis $a^p b^q$. In like manner if you consider the event as happening p times and failing q times in any other particular order, the probability for it is $a^p b^q$; but the number of different orders according to which an event may happen or fail, so as in all to happen p times and fail q, in $p + q$ trials is equal to the number of permutations that *aaaa bbb* admit of when the number of *a*'s is p, and the number of *b*'s is q. And this number is equal to E, the coefficient of the term in which occurs $a^p b^q$ when $(a + b)^{p+q}$ is expanded. The event therefore may happen p times and fail q in $p + q$ trials E different ways and no more, and its happening and failing these several different ways are so many inconsistent events, the probability for each of which is $a^p b^q$, and therefore by prop. 1 the probability that some way or other it happens p times and fails q times in $p + q$ trials is $Ea^p b^q$.

Section II

POSTULATE 1. I suppose the square table or plane $ABCD$ to be so made and levelled, that if either of the balls O or W be thrown upon it, there shall be the same probability that it rests upon any one equal part of the plane as another, and that it must necessarily rest somewhere upon it.

2. I suppose that the ball W shall be first thrown, and through the point where it rests a line os shall be drawn parallel to AD, and meeting CD and AB in s and o; and that afterwards the ball O shall be thrown $p + q$ or n times, and that its resting between AD and os after a single throw be called the happening of the event M in a single trial. These things supposed:

LEM. 1. The probability that the point o will fall between any two points in the line AB is the ratio of the distance between the two points to the whole line AB.

Let any two points be named, as f and b in the line AB, and through them parallel to AD draw fF, bL meeting CD in F and L. Then if the rectangles Cf, Fb, LA are commensurable to each other, they may each be divided into the same equal parts, which being done, and the ball W thrown, the probability it will rest somewhere upon any number of these equal parts will be the sum of the probabilities it has to rest upon each one of them, because its resting upon

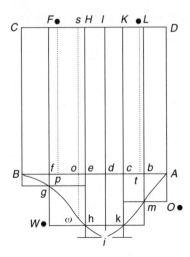

any different parts of the plane *AC* are so many inconsistent events; and this sum, because the probability it should rest upon any one equal part as another is the same, is the probability it should rest upon any one equal part multiplied by the number of parts. Consequently, the probability there is that the ball *W* should rest somewhere upon *Fb* is the probability it has to rest upon one equal part multiplied by the number of equal parts in *Fb*; and the probability it rests somewhere upon *Cf* or *LA*, i.e. that it does not rest upon *Fb* (because it must rest somewhere upon *AC*) is the probability it rests upon one equal part multiplied by the number of equal parts in *Cf*, *LA* taken together. Wherefore, the probability it rests upon *Fb* is to the probability it does not as the number of equal parts in *Fb* is to the number of equal parts in *Cf*, *LA* together, or as *Fb* to *Cf*, *LA* together, or as *fb* to *Bf*, *Ab* together. Wherefore the probability it rests upon *Fb* is to the probability it does not as *fb* to *Bf*, *Ab* together. And (*componendo inverse*) the probability it rests upon *Fb* is to the probability it rests upon *Fb* added to the probability it does not, as *fb* to *AB*, or as the ratio of *fb* to *AB* to the ratio of *AB* to *AB*. But the probability of any event added to the probability of its failure is the ratio of equality; wherefore, the probability it rests upon *Fb* is to the ratio of equality as the ratio of *fb* to *AB* to the ratio of *AB* to *AB*, or the ratio of equality; and therefore the probability it rests upon *Fb* is the ratio of *fb* to *AB*. But *ex hypothesi* according as the ball *W* falls upon *Fb* or not the point *o* will lie between *f* and *b* or not, and therefore the probability the point *o* will lie between *f* and *b* is the ratio of *fb* to *AB*.

Again; if the rectangles *Cf*, *Fb*, *LA* are not commensurable, yet the last mentioned probability can be neither greater nor less than the ratio of *fb* to *AB*; for, if it be less, let it be the ratio of *fc* to *AB*, and upon the line *fb* take the points *p* and *t*, so that *pt* shall be greater than *fc*, and the three lines *Bp*, *pt*, *tA* commensurable (which it is evident may be always done by dividing *AB* into equal parts less than half *cb*, and taking *p* and *t* the nearest points of division to *f* and *c* that lie upon *fb*). Then because *Bp*, *pt*, *tA* are commensurable, so are the rectangles *Cp*, *Dt*, and that upon *pt* compleating the square *AB*. Wherefore, by what has been said, the probability that the point *o* will lie between *p* and *t* is the ratio of *pt* to *AB*. But if it lies between *p* and *t* it must lie between *f* and *b*. Wherefore, the probability it should lie between *f* and *b* cannot be less than the ratio of *pt* to *AB*, and therefore must be greater than the ratio of *fc* to *AB* (since *pt* is greater than *fc*). And after the same manner you may prove that the forementioned probability cannot be greater than the ratio of *fb* to *AB*, it must therefore be the same.

LEM. 2. The ball *W* having been thrown, and the line *os* drawn, the probability of the event *M* in a single trial is the ratio of *Ao* to *AB*.

For, in the same manner as in the foregoing lemma, the probability that the ball *O* being thrown shall rest somewhere upon *Do* or between *AD* and *so* is the ratio of *Ao* to *AB*. But the resting of the ball *O* between *AD* and *so* after a single throw is the happening of the event *M* in a single trial. Wherefore the lemma is manifest.

Prop. 8

If upon *BA* you erect the figure *BghikmA* whose property is this, that (the base *BA* being divided into any two parts, as *Ab*, and *Bb* and at the point of division *b* a perpendicular being erected and terminated by the figure in *m*; and *y*, *x*, *r* representing respectively the ratio of *bm*, *Ab*, and *Bb* to *AB*, and *E* being the coefficient of the term in which occurs $a^p b^q$ when the binomial $(a + b)^{p+q}$ is expanded) $y = Ex^p r^q$. I say that before the ball *W* is thrown, the probability the point *o* should fall between *f* and *b*, any two points named in the line *AB*, and withal that the event *M* should happen *p* times and fail *q* in *a + b* trials, is the ratio of *fghikmb*, the part of the figure *BghikmA* intercepted between the perpendiculars *fg*, *bm* raised upon the line *AB*, to *CA* the square upon *AB*.

DEMONSTRATION

For if not; first let it be the ratio of *D* a figure greater than *fghikmb* to *CA*, and through the points *e*, *d*, *c* draw perpendiculars to *fb* meeting the curve *AmigB* in *h*, *i*, *k*; the point *d* being so placed that *di* shall be the longest of the perpendiculars terminated by the line *fb*, and the curve *AmigB*; and the points *e*, *d*, *c* being so many and so placed that the rectangles, *bk*, *ci*, *ei*, *fh* taken together shall differ less from *fghikmb* than *D* does; all which may be easily done by the help of the equation of the curve, and the difference between D and the figure *fghikmb* given. Then since *di* is the longest of the perpendicular ordinates that insist upon *fb*, the rest will gradually decrease as they are farther and farther from it on each side, as appears from the construction of the figure, and consequently *eh* is greater than *gf* or any other ordinate that insists upon *ef*.

Now if *Ao* were equal to *Ae*, then by lem. 2 the probability of the event M in a single trial would be the ratio of *Ae* to *AB*, and consequently by cor. Prop. 1 the probability of its failure would be the ratio of *Be* to *AB*. Wherefore, if x and r be the two forementioned ratios respectively, by Prop. 7 the probability of the event M happening p times and failing q in $p + q$ trials would be $Ex^p r^q$. But x and r being respectively the ratios of *Ae* to *AB* and *Be* to *AB*, if y is the ratio of *eh* to *AB*, then, by construction of the figure *AiB*, $y = Ex^p r^q$. Wherefore, if *Ao* were equal to *Ae* the probability of the event M happening p times and failing q in $p + q$ trials would be y, or the ratio of *eh* to *AB*. And if *Ao* were equal to *Af*, or were any mean between *Ae* and *Af*, the last mentioned probability for the same reasons would be the ratio of *fg* or some other of the ordinates insisting upon *ef*, to *AB*. But *eh* is the greatest of all the ordinates that insist upon *ef*. Wherefore, upon supposition the point should lie anywhere between *f* and *e*, the probability that the event M happens p times and fails q in $a + b$ trials cannot be greater than the ratio of *eh* to *AB*. There then being these two subsequent events, the 1st that the point *o* will lie between *e* and *f*, the 2nd that the event M will happen p times and fail q in $p + q$ trials, and the probability of the first (by lemma 1) is the ratio of *ef* to *AB*, and upon supposition the 1st happens, by what has been now proved, the probability of the 2nd cannot be greater than the ratio of *eh* to *AB*, it evidently follows (from Prop. 3) that the probability both together will happen cannot be greater than the ratio compounded of that of *ef* to *AB* and that of *eh* to *AB*, which compound ratio is the ratio of *fh* to *CA*. Wherefore, the probability that the point *o* will lie between *f* and *e*, and the event M happen p times and fail q, is not greater than the ratio of *fh* to *CA*. And in like manner the probability the point *o* will lie between *e* and *d*, and the event M happen and fail as before, cannot be greater than the ratio of *ei* to *CA*. And again, the probability the point *o* will lie between *d* and *c*, and the event M happen and fail as before, cannot be greater than the ratio of *ci* to *CA*. And lastly, the probability that the point *o* will lie between *c* and *b*, and the event M happen and fail as before, cannot be greater than the ratio of *bk* to *CA*. Add now all these several probabilities together, and their sum (by Prop. 1) will be the probability that the point will lie somewhere between *f* and *b*, and the event M happen p times and fail q in $p + q$ trials. Add likewise the correspondent ratios together, and their sum will be the ratio of the sum of the antecedents to their common consequent, i.e. the ratio of *fh*, *ei*, *ci*, *bk* together to *CA*; which ratio is less than that of *D* to *CA*, because *D* is greater than *fh, ei, ci, bk* together. And therefore, the probability that the point *o* will lie between *f* and *b*, and withal that the event *M* will happen *p* times and fail *q* in *p* + *q* trials, is less than the ratio of *D* to *CA*; but it was supposed the same which is absurd. And in like manner, by inscribing rectangles within the figure, as *eg, dh, dk, cm*, you may prove that the last mentioned probability is *greater* than the ratio of any figure less than *fghikmb* to *CA*.

Wherefore, that probability must be the ratio of *fghikmb* to *CA*.

Cor. Before the ball *W* is thrown the probability that the point *o* will lie somewhere between *A* and *B*, or somewhere upon the line *AB*, and withal that the event *M* will happen *p* times, and fail *p* in *p* + *q* trials is the ratio of the whole figure *AiB* to *CA*. But it is certain that the point *o* will lie somewhere upon *AB*. Wherefore, before the ball *W* is thrown the probability the event *M* will happen *p* times and fail *q* in *p* + *q* trials is the ratio of *AiB* to *CA*.

Prop. 9

If before anything is discovered concerning the place of the point *o*, it should appear that the event *M* had happened *p* times and failed *q* in *p* + *q* trials, and from hence I guess that the point *o* lies between any two points in the line *AB*, as *f* and *b*, and consequently that the probability of the event *M* in a single trial was somewhere between the ratio of *Ab* to *AB* and that of *Af* to *AB*: the probability I am in the right is the ratio of that part of the figure *AiB* described as before which is intercepted between perpendiculars erected upon *AB* at the points *f* and *b*, to the whole figure *AiB*.

For, there being these two subsequent events, the first that the point *o* will lie between *f* and *b*; the second that the event *M* should happen *p* times and fail *q* in *p* + *q* trials; and (by cor. prop. 8) the original probability of the second is the ratio of *AiB* to *CA*, and (by prop. 8) the probability of both is the ratio of *fghimb* to *CA*; wherefore (by prop. 5) it being first discovered that the second has happened, and from hence I guess that the first has happened also, the probability I am in the right is the ratio of *fghimb* to *AiB*, the point which was to be proved.

Cor. The same things supposed, if I guess that the probability of the event *M* lies somewhere between 0 and the ratio of *Ab* to *AB*, my chance to be in the right is the ratio of *Abm* to *AiB*.

Scholium

From the preceding proposition it is plain, that in the case of such an event as I there call *M*, from the number of times it happens and fails in a certain number of trials, without knowing anything more concerning it, one may give a guess whereabouts its probability is, and, by the usual methods computing the magnitudes of the areas there mentioned, see the chance that the guess is right. And that the same rule is the proper one to be used in the case of an event concerning the probability of which we absolutely know nothing antecedently to any trials made concerning it, seems to appear from the following consideration; viz. that concerning such an event I have no reason to think that, in a certain number of trials, it should rather happen any one possible number of times than another. For, on this account, I may justly reason concerning it as if its probability had been at first unfixed, and then determined in such a manner as to give me no reason to think that in a certain number of trials, it should rather happen any one possible number of times than another. But this is exactly the case of the event M. For before the ball W is thrown, which determines its probability in a single trial (by cor. prop. 8), the probability it has to happen p times and fail q in $p + q$ or n trials is the ratio of AiB to CA, which ratio is the same when $p + q$ or n is given, whatever number p is; as will appear by computing the magnitude of AiB by the method of fluxions.* And consequently before the place of the point o is discovered or the number of times the event M has happened in n trials, I can have no reason to think it should rather happen one possible number of times than another.

In what follows therefore I shall take for granted that the rule given concerning the event M in prop. 9 is also the rule to be used in relation to any event concerning the probability of which nothing at all is known antecedently to any trials made or observed concerning it. And such an event I shall call an unknown event.

COR. Hence, by supposing the ordinates in the figure AiB to be contracted in the ratio of E to one, which makes no alteration in the proportion of the parts of the figure intercepted between them, and applying what is said of the event M to an unknown event, we have the following proposition, which gives the rules for finding the probability of an event from the number of times it actually happens and fails.

Prop. 10

If a figure be described upon any base AH (Vid. Fig.) having for its equation $y = x^p r^q$; where y, x, r are respectively the ratios of an ordinate of the

---

* It will be proved presently in art. 4 by computing in the method here mentioned that AiB contracted in the ratio of E to 1 is to CA as 1 to $(n + 1)E$: from whence it plainly follows that, antecedently to this contraction, AiB must be to CA in the ratio of 1 to $n + 1$, which is a constant ratio when n is given, whatever p is.

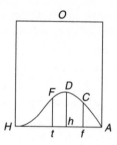

figure insisting on the base at right angles, of the segment of the base intercepted between the ordinate and *A* the beginning of the base, and of the other
segment of the base lying between the ordinate and the point *H*, to the base
as their common consequent. I say then that if an unknown event has
happened *p* times and failed *q* in *p* + *q* trials, and in the base *AH* taking any
two points as *f* and *t* you erect the ordinates *fC*, *tF* at right angles with it, the
chance that the probability of the event lies somewhere between the ratio of
Af to *AH* and that of *At* to *AH*, is the ratio of *tFCf*, that part of the before-
described figure which is intercepted between the two ordinates, to *ACFH* the
whole figure insisting on the base *AH*.

This is evident from prop. 9 and the remarks made in the foregoing
scholium and corollary.

Now, in order to reduce the foregoing rule to practice, we must find the
value of the area of the figure described and the several parts of it separated,
by ordinates perpendicular to its base. For which purpose, suppose *AH* = 1
and *HO* the square upon *AH* likewise = 1, and *Cf* will be = *y*, and *Af* = *x*, and
Hf = *r*, because *y*, *x* and *r* denote the ratios of *Cf*, *Af*, and *Hf* respectively to
AH. And by the equation of the curve $y = x^p r^q$ and (because *Af* + *fH* = *AH*)
r + *x* = 1. Wherefore

$$y = x^p(1 - x)^q$$

$$= x^p - qx^{p+1} + \frac{q(q - 1)x^{p+2}}{2} - \frac{q(q - 1)(q - 2)x^{p+3}}{2.3} + \text{etc.}$$

Now the abscisse being *x* and the ordinate x^p the correspondent area is $x^{p+1}/$
$(p + 1)$ (by prop. 10, cas. 1, Quadrat. Newt.)* and the ordinate being qx^{p+1} the

---

* 'Tis very evident here, without having recourse to Sir Isaac Newton, that the fluxion of the area *ACf*
being

$$\dot{y}x = x^p\dot{x} - qx^{p+1}\dot{x} + \frac{q(q - 1)}{2}x^{p+2}\dot{x} - \text{etc.}$$

the fluent or area itself is $\dfrac{x^{p+1}}{p + 1} - \dfrac{qx^{p+2}}{p + 2} + \dfrac{q(q - 1)x^{p+3}}{2(p + 3)} - \text{etc.}$

area is $qx^{p+2}/(p+2)$; and in like manner of the rest. Wherefore, the abscisse being x and the ordinate y or $x^p - qx^{p+1} +$ etc. the correspondent area is

$$\frac{x^{p+1}}{p+1} - \frac{qx^{p+2}}{p+2} + \frac{q(q-1)x^{p+3}}{2(p+3)} - \frac{q(q-1)(q-2)x^{p+4}}{2 \cdot 3(p+4)} + \text{etc.}$$

Wherefore, if $x = Af = Af/(AH)$, and $y = Cf = Cf/(AH)$, then

$$ACf = \frac{ACf}{HO} = \frac{x^{p+1}}{p+1} - \frac{qx^{p+2}}{p+2} + \frac{q(q-1)x^{p+3}}{2(p+3)} - \text{etc.}$$

From which equation, if q be a small number, it is easy to find the value of the ratio of ACf to HO and in like manner as that was found out, it will appear that the ratio of HCf to HO is

$$\frac{r^{q+1}}{q+1} - \frac{pr^{q+2}}{q+2} + \frac{p(p-1)r^{q+3}}{2(q+3)} - \frac{p(p-1)(p-2)r^{q+4}}{2.3(q+4)} + \text{etc.}$$

which series will consist of few terms and therefore is to be used when p is small.

2. The same things supposed as before, the ratio of ACf to HO is

$$\frac{x^{p+1}r^q}{p+1} + \frac{qx^{p+2}r^{q-1}}{(p+1)(p+2)} + \frac{q(q-1)x^{p+3}r^{q-2}}{(p+1)(p+2)(p+3)}$$

$$+ \frac{q(q-1)(q-2)x^{p+4}r^{q-3}}{(p+1)(p+2)(p+3)(p+4)} + \text{etc.} + \frac{x^{n+1}q(q-1)\dots 1}{(n+1)(p+1)(p+2)\dots n},$$

where $n = p + q$. For this series is the same with $x^{p+1}/(p+1) - qx^{p+2}/(p+2) + $ etc. set down in Art. 1st as the value of the ratio of ACf to HO; as will easily be seen by putting in the former instead of r its value $1 - x$, and expanding the terms and ordering them according to the powers of x. Or, more readily, by comparing the fluxions of the two series, and in the former instead of $\dot{r}$ substituting $-\dot{x}$.*

---

* The fluxion of the first series is

$$x^p r^q \dot{x} + \frac{qx^{p+1}r^{q-1}\dot{r}}{p+1} + \frac{qx^{p+1}r^{q-1}\dot{x}}{p+1} + \frac{q(q-1)x^{p+2}r^{q-2}\dot{r}}{(p+1)(p+2)}$$

$$+ \frac{q(q-1)x^{p+2}r^{q-2}\dot{x}}{(p+1)(p+2)} + \frac{q(q-1)(q-2)x^{p+3}r^{q-3}\dot{r}}{(p+1)(p+2)(p+3)} + \text{etc.}$$

or, substituting $-\dot{x}$ for $\dot{r}$,

$$x^p r^q \dot{x} - \frac{qx^{p+1}r^{q-1}\dot{x}}{p+1} + \frac{qx^{p+1}r^{q-1}\dot{x}}{p+1} - \frac{q(q-1)x^{p+2}r^{q-2}\dot{x}}{(p+1)(p+2)} + \frac{q(q-1)x^{p+2}r^{q-2}\dot{x}}{(p+1)(p+2)} - \text{etc.}$$

3. In like manner, the ratio of *HCf* to *HO* is

$$\frac{r^{q+1}x^p}{q+1} + \frac{pr^{q+2}x^{p-1}}{(q+1)(q+2)} + \frac{p(p-1)r^{q+3}x^{p-2}}{(q+1)(q+2)(q+3)} + \text{etc.}$$

4. If *E* be the coefficient of that term of the binomial $(a+b)^{p+q}$ expanded in which occurs $a^p b^q$, the ratio of the whole figure *ACFH* to *HO* is $\{(n+1)\,E\}^{-1}$, *n* being $= p + q$. For, when $Af = AH, x = 1, r = 0$. Wherefore, all the terms of the series set down in Art. 2 as expressing the ratio of *ACf* to *HO* will vanish except the last, and that becomes

$$\frac{q(q-1)\dots 1}{(n+1)(p+1)(p+2)\dots n}.$$

But *E* being the coefficient of that term in the binomial $(a+b)^n$ expanded in which occurs $a^p b^q$ is equal to

$$\frac{(p+1)(p+2)\dots n}{q(q-1)\dots 1}.$$

And, because *Af* is supposed to become $= AH, ACf = ACH$. From whence this article is plain.

5. The ratio of *ACf* to the whole figure *ACFH* is (by Art. 1 and 4)

$$(n+1)E\left[\frac{x^{p+1}}{p+1} - \frac{qx^{p+2}}{p+2} + \frac{q(q-1)x^{p+3}}{2(p+3)} - \text{etc.}\right]$$

and if, as *x* expresses the ratio of *Af* to *AH*, *X* should express the ratio of *At* to *AH*; the ratio of *AFt* to *ACFH* would be

$$(n+1)E\left[\frac{X^{p+1}}{p+1} - \frac{qX^{p+2}}{p+2} + \frac{q(q-1)X^{p+3}}{2(p+3)} - \text{etc.}\right]$$

which, as all the terms after the first destroy one another, is equal to

$$x^p r^q \dot{x} = x^p(1-x)^q \dot{x} = x^p \dot{x}\left[1 - qx + q\frac{(q-1)}{2}x^2 - \text{etc.}\right]$$

$$= x^p \dot{x} - qx^{p+1}\dot{x} + \frac{q(q-1)x^{p+2}}{2}\dot{x} - \text{etc.}$$

$$= \text{the fluxion of the latter series, or of } \frac{x^{p+1}}{p+1} - \frac{qx^{p+2}}{p+2} + \text{etc.}$$

The two series therefore are the same.

and consequently the ratio of *tFCf* to *ACFH* is $(n + 1)E$ multiplied into the difference between the two series. Compare this with prop. 10 and we shall have the following practical rule.

Rule 1

If nothing is known concerning an event but that it has happened p times and failed q in $p + q$ or n trials, and from hence I guess that the probability of its happening in a single trial lies somewhere between any two degrees of probability as X and x, the chance I am in the right in my guess is $(n + 1)E$ multiplied into the difference between the series

$$\frac{X^{p+1}}{p+1} - \frac{qX^{p+2}}{p+2} + \frac{q(q-1)X^{p+3}}{2(p+3)} - \text{etc.}$$

and the series

$$\frac{x^{p+1}}{p+1} - \frac{qx^{p+2}}{p+2} + \frac{q(q-1)x^{p+3}}{2(p+3)} - \text{etc.}$$

E being the coefficient of $a^p b^q$ when $(a + b)^n$ is expanded.

This is the proper rule to be used when q is a small number; but if q is large and p small, change everywhere in the series here set down p into q and q into p and x into r or $(1 - x)$, and X into $R = (1 - X)$; which will not make any alteration in the difference between the two series.

Thus far Mr Bayes's essay.

With respect to the rule here given, it is further to be observed, that when both p and q are very large numbers, it will not be possible to apply it to practice on account of the multitude of terms which the series in it will contain. Mr Bayes, therefore, by an investigation which it would be too tedious to give here, has deduced from this rule another, which is as follows.

Rule 2

If nothing is known concerning an event but that it has happened p times and failed q in $p + q$ or n trials, and from hence I guess that the probability of its happening in a single trial lies between $(p/n) + z$ and $(p/n) - z$; if $m^2 = n^3/(pq)$,* $a = p/n$, $b = q/n$, E the coefficient of the term in which occurs $a^p b^q$ when $(a + b)^n$ is expanded, and

$$\Sigma = \frac{(n + 1)\sqrt{(2pq)}}{n\sqrt{n}} Ea^p b^q$$

---

* [This equation is corrected to $m^2 = n^3/(2pq)$ by Price in his 1765 *Supplement*, see p. 133. Ed.]

multiplied by the series

$$mz - \frac{m^3z^3}{3} + \frac{(n-2)m^5z^5}{2n.5} - \frac{(n-2)(n-4)m^7z^7}{2n.3n.7}$$

$$+ \frac{(n-2)(n-4)(n-6)m^9z^9}{2n.3n.4n.9} - \text{etc.}$$

my chance to be in the right is greater than

$$\frac{2\Sigma*}{1 + 2Ea^pb^q + 2Ea^pb^q/n}$$

and less than

$$\frac{2\Sigma}{1 - 2Ea^pb^q - 2Ea^pb^q/n},$$

and if $p = q$ my chance is 2Σ exactly.

In order to render this rule fit for use in all cases it is only necessary to know how to find within sufficient nearness the value of Ea^pb^q and also of the series $mz - \frac{1}{3}m^3z^3 + $ etc. With respect to the former Mr Bayes has proved that, supposing K to signify the ratio of the quadrantal arc to its radius, Ea^pb^q will be equal to $\frac{1}{2}\sqrt{n}/\sqrt{(Kpq)}$ multiplied by the *ratio*, [h], whose *hyberbolic* logarithm is

$$\frac{1}{12}\left[\frac{1}{n} - \frac{1}{p} - \frac{1}{q}\right] - \frac{1}{360}\left[\frac{1}{n^3} - \frac{1}{p^3} - \frac{1}{q^3}\right] + \frac{1}{1260}\left[\frac{1}{n^5} - \frac{1}{p^5} - \frac{1}{q^5}\right]$$

$$- \frac{1}{1680}\left[\frac{1}{n^7} - \frac{1}{p^7} - \frac{1}{q^7}\right] + \frac{1}{1188}\left[\frac{1}{n^9} - \frac{1}{p^9} - \frac{1}{q^9}\right] - \text{etc.†}$$

where the numeral coefficients may be found in the following manner. Call them A, B, C, D, E etc. Then

$$A = \frac{1}{2.2.3} = \frac{1}{3.4}, \quad B = \frac{1}{2.4.5} - \frac{A}{3}, \quad C = \frac{1}{2.6.7} - \frac{10B + A}{5},$$

$$D = \frac{1}{2.8.9} - \frac{35C + 21B + A}{7}, \quad E = \frac{1}{2.10.11} - \frac{126C + 84D + 36B + A}{9},$$

---

* In Mr Bayes's manuscript this chance is made to be greater than $2\Sigma/(1 + 2Ea^pb^q)$ and less than $2\Sigma/(1 - 2Ea^pb^q)$. The third term in the two divisors, as I have given them, being omitted. But this being evidently owing to a small oversight in the deduction of this rule, which I have reason to think Mr Bayes had himself discovered, I have ventured to correct his copy, and to give the rule as I am satisfied it ought to be given.

† A very few terms of this series will generally give the hyperbolic logarithm to a sufficient degree of exactness. A similar series has been given by Mr De Moivre, Mr Simpson and other eminent

$$F = \frac{1}{2 . 12 . 13} - \frac{462D + 330C + 165E + 55B + A}{11} \text{ etc.}$$

where the coefficients of B, C, D, E, F, etc. in the values of D, E, F, etc. are the 2, 3, 4, etc. highest coefficients in $(a+b)^7$, $(a+b)^9$, $(a+b)^{11}$, etc. expanded; affixing in every particular value the least of these coefficients to B, the next in magnitude to the furthest letter from B, the next to C, the next to the furthest but one, the next to D, the next to the furthest but two, and so on.*

With respect to the value of the series

$$mz - \frac{1}{3}m^3z^3 + \frac{(n-2)m^5z^5}{2n . 5} \text{ etc.}$$

he has observed that it may be calculated directly when mz is less than 1, or even not greater than $\sqrt{3}$: but when mz is much larger it becomes impracticable to do this; in which case he shews a way of easily finding two values of it very nearly equal between which its true value must lie.

The theorem he gives for this purpose is as follows.

Let K, as before, stand for the ratio of the quadrantal arc to its radius, and H for the ratio whose hyperbolic logarithm is

$$\frac{2^2 - 1}{2n} - \frac{2^4 - 1}{360n^3} + \frac{2^6 - 1}{1260n^5} - \frac{2^8 - 1}{1680n^7} + \text{ etc.}$$

Then the series $mz - \frac{1}{3}m^3z^3 + \text{ etc.}$ will be greater or less than the series

$$\frac{Hn\sqrt{K}}{(n+1)\sqrt{2}} - \frac{n\left(1 - \frac{2m^2z^2}{n}\right)^{(1/2)n+1}}{(n+2)2mz} + \frac{n^2\left(1 - \frac{2m^2z^2}{n}\right)^{(1/2)n+2}}{(n+2)(n+4)4m^3z^3}$$

$$- \frac{3n^3\left(1 - \frac{2m^2z^2}{n}\right)^{(1/2)n+3}}{(n+2)(n+4)(n+6)8m^5z^5} + \frac{3 . 5 . n^4\left(1 - \frac{2m^2z^2}{n}\right)^{(1/2)n+4}}{(n+2)(n+4)(n+6)(n+8)16m^7z^7} - \text{ etc.}$$

mathematicians in an expression for the sum of the logarithms of the numbers 1, 2, 3, 4, 5, to x, which sum they have asserted to be equal to

$$\frac{1}{2}\log c + \left(x + \frac{1}{2}\right)\log x - x + \frac{1}{12x} - \frac{1}{360x^3} + \frac{1}{1260x^5} - \text{ etc.}$$

c denoting the circumference of a circle whose radius is unity. But Mr Bayes, in a preceding paper in this volume, has demonstrated that, though this expression will very nearly approach to the value of this sum when only a proper number of the first terms is taken, the whole series cannot express any quantity at all, because, let x be what it will, there will always be a part of the series where it will begin to diverge. This observation, though it does not much affect the use of this series, seems well worth the notice of mathematicians.

* This method of finding these coefficients I have deduced from the demonstration of the third lemma at the end of Mr Simpson's *Treatise on the Nature and Laws of Chance*.

continued to any number of terms, according as the last term has a positive or a negative sign before it.

From substituting these values of $Ea^p b^q$ and

$$mz - \frac{m^3 z^3}{3} + \frac{(n-2)\, m^5 z^5}{2n \cdot 5} \text{ etc.}$$

in the second rule arises a third rule, which is the rule to be used when mz is of some considerable magnitude.

Rule 3

If nothing is known of an event but that it has happened p times and failed q in $p + q$ or n trials, and from hence I judge that the probability of its happening in a single trial lies between $p/n + z$ and $p/n - z$ my chance to be right is *greater* than

$$\frac{(1/2)\sqrt{(Kpq)h}}{\sqrt{(Kpq)} + hn^{(1/2)} + hn^{-(1/2)}} \left\{ 2H - \frac{\sqrt{2(n+1)}\,(1 - 2m^2 z^2/n)^{(1/2)n+1}}{\sqrt{K}(n+2)mz} \right\}$$

and *less* than

$$\frac{(1/2)\sqrt{(Kpq)h}}{\sqrt{(Kpq)} - hn^{(1/2)} - hn^{-(1/2)}} \left\{ 2H - \frac{\sqrt{2(n+1)}\,(1 - 2m^2 z^2/n)^{(1/2)n+1}}{\sqrt{K}(n+2)mz} \right.$$
$$\left. + \frac{\sqrt{2n(n+1)}\,(1 - 2m^2 z^2/n)^{(1/2)n+2}}{\sqrt{K}(n+2)\,(n+4)2m^3 z^3} \right\}$$

where m^2, K, h and H stand for the quantities already explained.

AN APPENDIX

Containing an application of the foregoing Rules to some particular Cases

The first rule gives a direct and perfect solution in all cases; and the two following rules are only particular methods of approximating to the solution given in the first rule, when the labour of applying it becomes too great.

The first rule may be used in all cases where either p or q are nothing or not large. The second rule may be used in all cases where mz is less than $\sqrt{3}$; and the third in all cases where $m^2 z^2$ is greater than 1 and less than $(1/2)n$, if n is an even number and very large. If n is not large this last rule cannot be much wanted, because, m decreasing continually as n is diminished, the value of z may in this case be taken large, (and therefore a considerable interval had between $p/n - z$ and $p/n + z$), and yet the operation be carried on by the second rule; or mz not exceed $\sqrt{3}$.

But in order to shew distinctly and fully the nature of the present problem, and how far Mr Bayes has carried the solution of it; I shall give the result of this solution in a few cases, beginning with the lowest and most simple.

Let us then first suppose, of such an event as that called *M* in the essay, or an event about the probability of which, antecedently to trials, we know nothing, that it has happened *once*, and that it is enquired what conclusion we may draw from hence with respect to the probability of its happening on a *second* trial.

The answer is that there would be an odds of three to one for somewhat more than an even chance that it would happen on a second trial.

For in this case, and in all others where *q* is nothing, the expression

$$(n + 1)\left\{\frac{X^{p+1}}{p + 1} - \frac{x^{p+1}}{p + 1}\right\} \quad \text{or} \quad X^{p+1} - x^{p+1}$$

gives the solution, as will appear from considering the first rule. Put therefore in this expression $p + 1 = 2$, $X = 1$ and $x - (1/2)$ and it will be $1 - (1/2)^2$ or 3/4; which shews the chance there is that the probability of an event that has happened once lies somewhere between 1 and 1/2; or (which is the same) the odds that it is somewhat more than an even chance that it will happen on a second trial.*

In the same manner it will appear that if the event has happened twice, the odds now mentioned will be seven to one; if thrice, fifteen to one; and in general, if the event has happened *p* times, there will be an odds of $2^{p-1} - 1$ to one, for *more* than an equal chance that it will happen on further trials.

Again, suppose all I know of an event to be that it has happened ten times without failing, and the enquiry to be what reason we shall have to think we are right if we guess that the probability of its happening in a single trial lies somewhere between 16/17 and 2/3, or that the ratio of the causes of its happening to those of its failure is some ratio between that of sixteen to one and two to one.

Here $p + 1 = 11$, $X = 16/17$ and $x = 2/3$ and $X^{p+1} - x^{p-1} = (16/17)^{11} - (2/3)^{11} = 0.5013$ etc. The answer therefore is, that we shall have very nearly an equal chance for being right.

In this manner we may determine in any case what conclusion we ought to draw from a given number of experiments which are unopposed by contrary experiments. Every one sees in general that there is reason to expect an event with more or less confidence according to the greater or less number of times in which, under given circumstances, it has happened without failing; but we here see exactly what this reason is, on what principles it is founded, and how we ought to regulate our expectations.

But it will be proper to dwell longer on this head.

Suppose a solid or die of whose number of sides and constitution we know nothing; and that we are to judge of these from experiments made in throwing it.

In this case, it should be observed, that it would be in the highest degree improbable that the solid should, in the first trial, turn any one side which could be assigned beforehand; because it would be known that some side it must turn, and that there was an infinity of other sides, or sides otherwise marked, which it was equally likely that it should turn. The first throw only shews that *it has* the side then thrown, without giving any reason to think that it has it any one number of times rather than any other. It will appear, therefore, that *after* the first throw and not before, we should be in the circumstances required by the conditions of the present problem, and that the whole effect of this throw would be to bring us into these circumstances. That is: the turning the side first thrown in any subsequent

---

* There can, I suppose, be no reason for observing that on this subject unity is always made to stand for certainty, and 1/2 for an even chance.

single trial would be an event about the probability or improbability of which we could form no judgment, and of which we should know no more than that it lay somewhere between nothing and certainty. With the second trial then our calculations must begin; and if in that trial the supposed solid turns again the same side, there will arise the probability of three to one that it has more of that sort of sides than of *all* others; or (which comes to the same) that there is somewhat in its constitution disposing it to turn that side oftenest: And this probability will increase, in the manner already explained, with the number of times in which that side has been thrown without failing. It should not, however, be imagined that any number of such experiments can give sufficient reason for thinking that it would *never* turn any other side. For, suppose it has turned the same side in every trial a million of times. In these circumstances there would be an improbability that it has *less* than 1,400,000 more of these sides than all others; but there would also be an improbability that it had *above* 1,600,000 times more. The chance for the latter is expressed by 1,600,000/1,600,001 raised to the millionth power subtracted from unity, which is equal to 0·4647 etc and the chance for the former is equal to 1,400,000/1,400,001 raised to the same power, or to 0·4895; which, being both less than an equal chance, proves what I have said. But though it would be thus improbable that it had *above* 1,600,000 times more or *less* than 1,400,000 times *more* of these sides than of all others, it by no means follows that we have any reason for judging that the true proportion in this case lies somewhere between that of 1,600,000 to one and 1,400,000 to one. For he that will take the pains to make the calculation will find that there is nearly the probability expressed by 0·527, or but little more than an equal chance, that it lies somewhere between that of 600,000 to one and three millions to one. It may deserve to be added, that it is more probable that this proportion lies somewhere between that of 900,000 to 1 and 1,900,000 to 1 than between any other two proportions whose antecedents are to one another as 900,000 to 1,900,000, and consequents unity.

I have made these observations chiefly because they are all strictly applicable to the events and appearances of nature. Antecedently to all experience, it would be improbable as infinite to one, that any particular event, beforehand imagined, should follow the application of any one natural object to another; because there would be an equal chance for any one of an infinity of other events. But if we had once seen any particular effects, as the burning of wood on putting it into fire, or the falling of a stone on detaching it from all contiguous objects, then the conclusions to be drawn from any number of subsequent events of the same kind would be to be determined in the same manner with the conclusions just mentioned relating to the constitution of the solid I have supposed. In other words. The first experiment supposed to be ever made on any natural object would only inform us of one event that may follow a particular change in the circumstances of those objects; but it would not suggest to us any ideas of uniformity in nature, or give us the least reason to apprehend that it was, in that instance or in any other, regular rather than irregular in its operations. But if the same event has followed without interruption in any one or more subsequent experiments, then some degree of uniformity will be observed; reason will be given to expect the same success in further experiments, and the calculations directed by the solution of this problem may be made.

One example here it will not be amiss to give.

Let us imagine to ourselves the case of a person just brought forth into this world, and left to collect from his observation of the order and course of events what powers and causes take place in it. The Sun would, probably, be the first object that would engage his attention; but after losing it the first night he would be entirely ignorant whether he should ever see it again. He would therefore be in the condition of a person making a first experiment about an event entirely unknown to him. But let him see a second appearance or one *return* of the Sun, and an expectation would be raised in him of a second return, and he might know that there was an odds of 3 to 1 for *some* probability of this. This odds would increase, as before represented, with the number of returns to which he was witness. But no finite number of returns would be sufficient to produce absolute or physical certainty. For let it be supposed that he has seen it return at regular and stated intervals a million of times. The conclusions this would warrant would be such as follow. There would be the odds of the millioneth power of 2, to one, that it was likely that it would return again at the end of the usual interval. There would be the probability expressed by 0·5352, that the odds for this was not *greater* than 1,600,000 to 1; and the probability expressed by 0·5105, that it was not less than 1,400,000 to 1.

It should be carefully remembered that these deductions suppose a previous total ignorance of nature. After having observed for some time the course of events it would be found that the operations of nature are in general regular, and that the powers and laws which prevail in it are stable and permanent. The consideration of this will cause one or a few experiments often to produce a much stronger expectation of success in further experiments than would otherwise have been reasonable; just as the frequent observation that things of a sort are disposed together in any place would lead us to conclude, upon discovering there any object of a particular sort, that there are laid up with it many others of the same sort. It is obvious that this, so far from contradicting the foregoing deductions, is only one particular case to which they are to be applied.

What has been said seems sufficient to shew us what conclusions to draw from *uniform* experience. It demonstrates, particularly, that instead of proving that events will *always* happen agreeably to it, there will be always reason against this conclusion. In other words, where the course of nature has been the most constant, we can have only reason to reckon upon a recurrency of events proportioned to the degree of this constancy; but we can have no reason for thinking that there are no causes in nature which will *ever* interfere with the operations of the causes from which this constancy is derived, or no circumstances of the world in which it will fail. And if this is true, supposing our only *data* derived from experience, we shall find additional reason for thinking thus if we apply other principles, or have recourse to such considerations as reason, independently of experience, can suggest.

But I have gone further than I intended here; and it is time to turn our thoughts to another branch of this subject: I mean, to cases where an experiment has sometimes succeeded and sometimes failed.

Here, again, in order to be as plain and explicit as possible, it will be proper to put the following case, which is the easiest and simplest I can think of.

Let us then imagine a person present at the drawing of a lottery, who knows nothing of its scheme or of the proportion of *Blanks* to *Prizes* in it. Let it further be supposed, that he is obliged to infer this from the number of *blanks* he hears drawn compared with the number of *prizes*; and that it is enquired what conclusions in these circumstances he may reasonably make.

Let him first hear *ten* blanks drawn and *one* prize, and let it be enquired what chance he will have for being right if he guesses that the proportion of *blanks* to *prizes* in the lottery lies somewhere between the proportions of 9 to 1 and 11 to 1.

Here taking $X = 11/12$, $x = 9/10$, $p = 10$, $q = 1$, $n = 11$, $E = 11$, the required chance, according to the first rule, is $(n + 1)E$ multiplied by the difference between

$$\left\{\frac{X^{p+1}}{p+1} - \frac{qX^{p+2}}{p+2}\right\} \quad \text{and}$$

$$\left\{\frac{x^{p+1}}{p+1} - \frac{qx^{p+2}}{p+2}\right\} = 12 \cdot 11 \cdot \left\{\left[\frac{(11/12)^{11}}{11} - \frac{(11/12)^{12}}{12}\right] - \left[\frac{(9/10)^{11}}{11} - \frac{(9/10)^{12}}{12}\right]\right\}$$

$$= 0 \cdot 07699 \text{ etc.}$$

There would therefore be an odds of about 923 to 76, or nearly 12 to 1 *against* his being right. Had he guessed only in general that there were less than 9 blanks to a prize, there would have been a probability of his being right equal to $0 \cdot 6589$, or the odds of 65 to 34.

Again, suppose that he has heard 20 *blanks* drawn and 2 *prizes*; what chance will he have for being right if he makes the same guess?

Here X and x being the same, we have $n = 22$, $p = 20$, $q = 2$, $E = 231$, and the required chance equal to

$$(n + 1)E\left\{\left[\frac{X^{p+1}}{p+1} - \frac{qX^{p+2}}{p+2} + \frac{q(q-1)X^{p+3}}{2(p+3)}\right] - \left[\frac{x^{p+1}}{p+1} - \frac{qx^{p+2}}{p+2} + \frac{q(q-1)x^{p+3}}{2(p+3)}\right]\right\}$$

$$= 0 \cdot 10843 \text{ etc.}$$

He will, therefore, have a better chance for being right than in the former instance, the odds against him now being 892 to 108 or about 9 to 1. But should he only guess in general, as before, that there were less than 9 blanks to a prize, his chance for being right will be worse; for instead of $0 \cdot 6589$ or an odds of near two to one, it will be $0 \cdot 584$, or an odds of 584 to 415.

Suppose, further, that he has heard 40 *blanks* drawn and 4 *prizes*; what will the before-mentioned chances be?

The answer here is $0 \cdot 1525$, for the former of these chances; and $0 \cdot 527$, for the latter. There will, therefore, now be an odds of only $5\frac{1}{2}$ to 1 against the proportion of blanks to prizes lying between 9 to 1 and 11 to 1; and but little more than an equal chance that it is less than 9 to 1.

Once more. Suppose he has heard 100 *blanks* drawn and 10 *prizes*.

The answer here may still be found by the first rule; and the chance for a proportion of blanks to prizes *less* than 9 to 1 will be $0 \cdot 44109$, and for a proportion *greater* than 11 to 1, $0 \cdot 3082$. It would therefore be likely that there were not *fewer* than 9 or *more* than 11 blanks to a prize. But at the same time it will remain unlikely* that the true proportion should lie between 9 to 1 and 11 to 1, the chance for this being $0 \cdot 2506$ etc. There will therefore be still an odds of near 3 to 1 against this.

From these calculations it appears that, in the circumstances I have supposed, the chance for being right in guessing the proportion of *blanks* to *prizes* to be nearly the same

---

* I suppose no attentive person will find any difficult in this. It is only saying that, supposing the interval between nothing and certainty divided into a hundred equal chances, there will be 44 of them for a less proportion of blanks to prizes than 9 to 1, 31 for a greater than 11 to 1, and 25 for some proportion between 9 to 1 and 11 to 1; in which it is obvious that, though one of these suppositions must be true, yet, having each of them more chances against them than for them, they are all separately unlikely.

with that of the number of *blanks* drawn in a given time to the number of prizes drawn, is continually increasing as these numbers increase; and that therefore, when they are considerably large, this conclusion may be looked upon as morally certain. By parity of reason, it follows universally, with respect to every event about which a great number of experiments has been made, that the causes of its happening bear the same proportion to the causes of its failing, with the number of happenings to the number of failures: and that, if an event whose causes are supposed to be known, happens oftener or seldomer than is agreeable to this conclusion, there will be reason to believe that there are some unknown causes which disturb the operations of the known ones. With respect, therefore, particularly to the course of events in nature, it appears, that there is demonstrative evidence to prove that they are derived from permanent causes, or laws originally established in the constitution of nature in order to produce that order of events which we observe, and not from any of the powers of chance.* This is just as evident as it would be, in the case I have insisted on, that the reason of drawing 10 times more *blanks* than *prizes* in millions of trials, was that there were in the wheel about so many more *blanks* than *prizes*.

But to proceed a little further in the demonstration of this point.

We have seen that supposing a person, ignorant of the whole scheme of a lottery, should be led to conjecture, from hearing 100 *blanks* and 10 prizes drawn, that the proportion of *blanks* to *prizes* in the lottery was somewhere between 9 to 1 and 11 to 1, the chance for his being right would be 0·2506 etc. Let [us] now enquire what this chance would be in some higher cases.

Let it be supposed that *blanks* have been drawn 1000 times, and prizes 100 times in 1100 trials.

In this case the powers of X and x rise so high, and the number of terms in the two series

$$\frac{X^{p+1}}{p+1} - \frac{qX^{p+2}}{p+2} \text{ etc. and } \frac{x^{p+1}}{p+1} - \frac{qx^{p+2}}{p+2} \text{ etc.}$$

become so numerous that it would require immense labour to obtain the answer by the first rule. 'Tis necessary, therefore, to have recourse to the second rule. But in order to make use of it, the interval between X and x must be a little altered. $(10/11) - (9/10)$ is $1/110$, and therefore the interval between $(10/11) - (1/110)$ and $(10/11) + (1/110)$ will be nearly the same with the interval between $9/10$ and $11/12$, only somewhat larger. If then we make the question to be; what chance there would be (supposing no more known than that blanks have been drawn 1000 times and prizes 100 times in 1100 trials) that the probability of drawing a blank in a single trial would lie somewhere between $(10/11) - (1/110)$ and $(10/11) + (1/110)$ we shall have a question of the same kind with the preceding questions, and deviate but little from the limits assigned in them.

The answer, according to the second rule, is that this chance is greater than

$$\frac{2\Sigma}{1 + 2Ea^p b^q + \dfrac{2Ea^p b^q}{n}}$$

and less than

$$\frac{2\Sigma}{1 - 2Ea^p b^q - 2E\dfrac{a^p b^q}{n}}$$

* See Mr De Moivre's *Doctrine of Chances*, p. 250.

Σ being $\dfrac{(n+1)\sqrt{(2pq)}}{n\sqrt{n}}\, Ea^p b^q \left\{ mz - \dfrac{m^3 z^3}{3} + \dfrac{(n-2)m^5 z^5}{2n\cdot 5} - \text{etc.} \right\}$

By making here $1000 = p$, $100 = q$, $1100 = n$, $1/110 = z$,

$$mz = z\sqrt{\left(\dfrac{n^3}{pq}\right)} = 1\cdot048808, \quad Ea^p b^q = \dfrac{1}{2}h\dfrac{\sqrt{n}}{\sqrt{(Kpq)}},$$

h being the ratio whose hyperbolic logarithm is

$$\dfrac{1}{12}\left[\dfrac{1}{n} - \dfrac{1}{p} - \dfrac{1}{q}\right] - \dfrac{1}{360}\left[\dfrac{1}{n^3} - \dfrac{1}{p^3} - \dfrac{1}{q^3}\right] + \dfrac{1}{1260}\left[\dfrac{1}{n^5} - \dfrac{1}{p^5} - \dfrac{1}{q^5}\right] - \text{etc.}$$

and K the ratio of the quadrantal arc to radius; the former of these expressions will be found to be 0·7953, and the latter 0·9405 etc. The chance enquired after, therefore, is greater than 0·7953, and than 0·9405. That is; there will be an odds for being right in guessing that the proportion of blanks to prizes lies *nearly* between 9 to 1 and 11 to 1, (or *exactly* between 9 to 1 and 1111 to 99), which is greater than 4 to 1, and less than 16 to 1.

Suppose, again, that no more is known than that *blanks* have been drawn 10,000 times and *prizes* 1000 times in 11,000 trials; what will the chance now mentioned be?

Here the second as well as the first rule becomes useless, the value of mz being so great as to render it scarcely possible to calculate directly the series

$$\left\{ mz - \dfrac{m^3 z^3}{3} + \dfrac{(n-2)m^5 z^5}{2n\cdot 5} - \text{etc.} \right\}$$

The third rule, therefore, must be used; and the information it gives us is, that the required chance is greater than 0·97421, or more than an odds of 40 to 1.

By calculations similar to these may be determined universally, what expectations are warranted by any experiments, according to the different number of times in which they have succeeded and failed; or what should be thought of the probability that any particular cause in nature, with which we have any acquaintance, will or will not, in any single trial, produce an effect that has been conjoined with it.

Most persons, probably, might expect that the chances in the specimen I have given would have been greater than I have found them. But this only shews how liable we are to error when we judge on this subject independently of calculation. One thing, however, should be remembered here; and that is, the narrowness of the interval between 9/10 and 11/12, or between $(10/11) + (1/110)$ and $(10/11) - (1/110)$. Had this interval been taken a little larger, there would have been a considerable difference in the results of the calculations. Thus had it been taken double, or $z = 1/55$, it would have been found in the fourth instance that instead of odds against there were odds for being right in judging that the probability of drawing a blank in a single trial lies between $(10/11) + (1/55)$ and $(10/11) - (1/55)$.

The foregoing calculations further shew us the uses and defects of the rules laid down in the essay. 'Tis evident that the two last rules do not give us the required chances within such narrow limits as could be wished. But here again it should be considered, that these limits become narrower and narrower as q is taken larger in respect of p; and when p and q are equal, the exact solution is given in all cases by the second rule. These two rules therefore afford a direction to our judgment that may be of considerable use till some person shall discover a better approximation to the value of the two series in the first rule.*

But what most of all recommends the solution in this *Essay* is, that it is compleat in those cases where information is most wanted, and where Mr De Moivre's solution of the inverse problem can give little or no direction; I mean, in all cases where either p or q are of no considerable magnitude. In other cases, or when both p and q are very considerable, it is not difficult to perceive the truth of what has been here demonstrated, or that there is reason to believe in general that the chances for the happening of an event are to the chances for its failure in the same *ratio* with that of p to q. But we shall be greatly deceived if we judge in this manner when either p or q are small. And tho' in such cases the *Data* are not sufficient to discover the exact probability of an event, yet it is very agreeable to be able to find the limits between which it is reasonable to think it must lie, and also to be able to determine the precise degree of assent which is due to any conclusions or assertions relating to them.

---

* Since this was written I have found out a method of considerably improving the approximation in the second and third rules by demonstrating that the expression $2\Sigma\{1 - 2Ea^p b^q - 2Ea^p b^q/n\}$ comes, in most cases, almost as near to the true value wanted as there is reason to desire, only always somewhat less. It seems necessary to hint this here: though the proof of it cannot be given.